Genre Change in the Contemporary World

Linguistic Insights

Studies in Language and Communication

Edited by Maurizio Gotti,
University of Bergamo

Volume 159

PETER LANG
Bern · Berlin · Bruxelles · Frankfurt am Main · New York · Oxford · Wien

Giuliana Garzone, Paola Catenaccio, Chiara Degano (eds)

Genre Change in the Contemporary World

Short-term Diachronic Perspectives

PETER LANG
Bern · Berlin · Bruxelles · Frankfurt am Main · New York · Oxford · Wien

Bibliographic information published by die Deutsche Nationalbibliothek
Die Deutsche Nationalbibliothek lists this publication in the Deutsche Nationalbibliografie; detailed bibliographic data is available on the Internet at ‹http://dnb.d-nb.de›.

British Library Cataloguing-in-Publication Data: A catalogue record for this book is available from The British Library, Great Britain

Library of Congress Cataloging-in-Publication Data

Genre change in the contemporary world : short-term diachronic perspectives / Giuliana Garzone, Paola Catenaccio, Chiara Degano (eds).
p. cm. – (Linguistic insights : Studies in language and communication; v. 159)
Includes bibliographical references.
ISBN 978-3-0343-1214-1
1. Discourse analysis. 2. Communication, International. 3. Communication and technology. 4. Popular culture and globalization. 5. Communication in organizations. 6. Interdisciplinary research. 7. Sublanguage. I. Garzone, G. (Giuliana) II. Catenaccio, Paola. III. Degano, Chiara.
P302.865.G46 2012
401'.41–dc23

2012024167

Published with a gr [illegible] inistry of Education, University and Research (National [illegible] *Tension and Change in Domain-specific Genres*, grant No. [illegible] the Dipartimento di Lingue e Culture Comtemporanee, [illegible] di Milano, Italy

ISSN 1424-8689
ISBN 978-3-0343-[illegible]

Hochfeldstrasse 32, CH-3012 Bern, Switzerland
info@peterlang.com, www.peterlang.com, www.peterlang.net

Printed in Switzerland

Contents

Institutional and Business Discourse

Giuliana Garzone / Paola Catenaccio / Chiara Degano

Introduction

Over the last few decades, rapid and extensive changes in communicative practices, patterns and technologies have deeply affected the generic configuration of professional and disciplinary domains, bringing about novel scenarios which call for new theoretical and analytical approaches. If some existing genres have survived the global digital revolution relatively unscathed – although frequently in a re-mediated form – others have undergone drastic changes, and still others have declined, while new ones have emerged. These developments call for a reconsideration of the repertoires of conventions traditionally identified in each specific genre as well as for a reassessment of the analytical tools used to investigate them.

The chapters of this volume focus on the evolution of genres in specialized communication under the pressure of technological innovations and the profound social changes triggered in the contemporary world by globalization. They also look at the possible implications of these evolutions for the methodology used in the analysis of specialised genres and its updating about three decades after the emergence of genre analysis.

The volume is opened by a chapter by Giuliana Garzone, which sets the stage for the following essays by providing an account of the forces and pressures underlying genre evolution. Her theory-informed discussion provides a backdrop against which to read the following chapters, which deal with specific aspects of genre evolution in different domains. Garzone aims to identify and systematize the main factors that have contributed to determining the extremely rapid process of genre change that has occurred in the last few decades, and the rise of new genres. The first change-inducing factors considered regard evolutions in professional and disciplinary genres originating within the various discourse communities. Here, a prominent role is played by interdiscursivity, i.e. the appropriation of semiotic resources across professional genres, practices and cultures, which determines changes due to contamination and overspill across discourse communities. Another important element is the more or less direct imposition of conventions in certain professional areas, cases in point being the gatekeeping action exerted by international journals on the genre of the scientific research article, and the imposition of new drafting conventions by governmental bodies in various Western countries, leading to a radical renewal in legislative genres. The other types of genre change considered do not originate 'locally', within the relevant discourse communities, but rather result from the impact of broader developments in society, and in particular from evolutions in predominant ethical and societal values, which often involve profound alterations in discursive practices and the need to rely on discourse formats that suit new exigencies, as has been the case for instance with emerging genres in the area of Corporate Social Responsibility and in green advertising. But according to Garzone the most radical transformations have occurred as a result of the spread of new technologies, mostly associated with Computer- and Web-Mediated Communication, which have opened up new, unexpected and virtually boundless discursive and social spaces, determining profound changes in existing genres as well as the rise of brand new genres. Overall, Garzone's chapter provides a broad picture of the factors determining genre change in the contemporary world, shedding light on the extremely complex mechanisms involved and providing categorizations that may be useful when reading the subsequent chapters.

After this opening chapter, the book is divided into two broad sections. The first one covers genre change and evolution in academic and scientific discourse, while the second one focuses on genres in institutional and business discourse. Opening the section devoted to academic and scientific discourse is MARINA BONDI and SILVIA CAVALIERI's analysis of the evolution of the genre of the academic abstract. Their study follows in the steps of a consolidated tradition of studies centred on empirical research-based genres (research articles, abstracts, etc.) often combining linguistic and rhetorical analysis, and examines markers of the 'representational' nature of abstracts in two comparable corpora collected respectively in 1988 and 2008 from seven refereed applied linguistics journals. In particular, it looks at the foregrounding of metadiscursive references to the original article and its procedures as objects of abstracting discourse, in relation to the type of article abstracted, observing rhetorical and lexical change. Results show that the development of the abstract seems to emphasize what Hyland (2000: 63) has referred to as promotional features of the genre. Another important finding is that impersonality – traditionally a distinctive feature of abstracts – seems to have given way to locativity in characterizing the writer's voice. Locational patterns allow writers a 'middle' way between personalization and impersonality: talking about their papers rather than about themselves they can be more explicit in the assessment of their own research, while at the same time retaining a somewhat more objective stance. Bondi and Cavalieri conclude that the combination of locational patterns and references to the community allows writers to strengthen their personal position by reference to collegiality, highlighting the role played by discursive markers of community membership in the evolution of the genre.

The role of the scientific community in shaping (and changing) academic genres and genre networks is also at the heart of DAVIDE GIANNONI's chapter, which examines the genre dynamics of a journal on the basis of variations in its tables of contents and texts over a 40-year period. The results indicate that generic diversity has been dramatically eroded over the years, with articles displacing almost all other contributions by the mid-1990s, and in particular the more dialogic/evaluative genres such as notes, comments and replies. The communicative purpose of these non-central genres is then investigated through keyword analysis, using a selection of the journal's entire repertoire as a reference corpus. Editorials are found to be the most marked for aboutness, comments and replies for stance/modality, and notes for style. The balancing of these three planes accounts for the distinctness of each genre, regardless of its length or range of topics. The overall impression points to a widening gap between articles and minor contributions that enrich the debate but are unlikely to be cited. Thus editorial decisions promote certain genres and phase out others in line with the argument that generic change within a discipline is largely dependent on the 'power and politics' of its institutional members.

The elements of diachronic variation identified by STEFANIA MACI in the third chapter in this section are determined both by needs inherent in the conventions of the discipline involved, and by external factors. Maci investigates the ways in which the impact of an evolving and increasingly business-oriented health care system has determined genre variation in medical discourse, focusing on posters presented at medical conferences over a span of time from the 1980s to the 2000s. Results indicate that the language of medical posters has gone from narration to a disjunctive mode which makes frequent recourse to bulleted lists similar to those used in promotional and advertising discourse, a feature obviously determined by the need to follow the Introduction-Methods-Results-Discussion pattern of scientific discourse in a genre that requires high levels of synthesis. Another important evolution detected is determined by external factors, as competition, financial constraints and medical worth have generated generic tensions within the medical discourse community that are apparently resolved through a marketization process. This is confirmed by the keyword lists generated from Maci's three corpora, which indicate that while in the years 1980-1999 attention was focused on statistical and empirical analysis, in the 2000-2009 decade there appears to be a more precise indication of credentials and a particular attention to *socio-economic, -political* and *-cultural* issues as fundamental factors in health care.

An element of marketisation also figures among the factors determining genre change in MARTIN SOLLY's chapter, which deals with an academic genre – commencement speeches – that is neither inter- nor intraspecialistic, but rather crosses over from the academia to the general public. Commencement speeches are delivered at graduation ceremonies in North American universities by prominent professional and academic figures, and over the years have become increasingly public moments, often widely reported in the press and available on the Internet. They are delivered in an academic setting and thus conform to the traditional packaging and format of the host institution. However, their construction and delivery are the responsibility of the speakers who have almost complete freedom as regards content, and are often expected to entertain the graduating students as well as provide words of wisdom and advice. In particular the chapter examines the rhetorical strategies and moves deployed in commencement speeches since 1947. The analysis shows that this genre evolution has been strongly influenced by the marketisation of higher education and by social change. These days the monologues are becoming increasingly dialogic, with speakers actively engaging their audiences and using a wide range of interactive language strategies. Despite this, some of the main characteristics of this genre, such as their traditional framing in the academic setting, have not changed at all over the years.

By virtue of their broadened participation framework and communicative purpose, commencement speeches bridge the gap between academia and the general public, a function typically ascribed to popularising genres, which are the topic of the last two chapters in this section. Both chapters deal with the popularisation of scientific discourse, respectively in Spain and in Taiwan. PAULA DE SANTIAGO explores the evolution of web genres in online newspapers, comparing scientific popularised articles (SPAs) published in the online supplement *Salud* and the online section *WebSalud* of the Spanish newspaper *El Mundo*. In particular, she focuses on the form and functionalities of the articles investigated, with additional insights drawn from a number of interviews conducted with the Health Department of the newspaper *El Mundo* in December 2009. The findings show that social relevance plays a key role in the selection of the topics covered, and suggest that the hypertextual and hypermodal affordances of the World Wide Web are exploited to a fuller extent in the articles that are designed for web-publication only, which are typically accompanied by hyperlinks to other articles on the same topic that help build a network of texts suitable for further exploration on the part of the reader.

The following chapter, by MIN-HSIU LIAO, focuses on the evolution of the popularising science article in Taiwan over the last thirty years of the twentieth century as testified by the changes undergone by *Science Monthly*, the most influential science magazine in the country. The study argues that one of the most powerful forces driving these changes is the relationship between expert scientists and lay readers. The data are drawn from a corpus of texts from the Q&A column in *Science Monthly*, spanning across three decades (1970-1999). To account for the entire range of interactions in the Q&A column, which involve multiple agents, including editors and readers, Liao's model of analysis is adopted from critical genre analysis as proposed by Bhatia (2004) and investigates discourse in three spaces: the textual space (between questioners and answerers), the institutional space (between readers and editors), and the social space (between lay public and expert scientists). The study concludes that features at the three levels interact with each other, not necessarily in a top-down or bottom-up direction, but rather through mutual influence and by shaping the relationship among genre users. Liao's chapter, which closes the section devoted to academic and scientific discourse, highlights the mutual interdependence of generic features, participant roles, and communicative intents.

These aspects are also of crucial importance for the studies included in the second section of the volume, which focuses on institutional and business communication, an area where the most important factors giving rise to genre evolution seem to be connected with the enlargement of the potential audience, also deriving from the online publication of documents. The section is opened by ANNARITA TAVANI's study of the evolution of the FAO report on the State of Food Insecurity (SOFI) from 1999 to 2009 as an example of a domain-specific genre developed within a community of practice consisting of policy makers, but also aimed at the public at large. Findings indicate that while the basic structure of the documents shows a rather high level of stability, a few significant changes have occurred over time. In particular, the distribution of text types varies in the several editions as does the overall organization of topics. Emphasis shifts from a description of the circumstances originating the SOFI reports to an analysis of the biomedical parameters underlying the study of Food Insecurity, while in later examples of the genre special attention is given to issues regarding developing countries. Changes in the use of microlinguistic features both at the level of lexis and at the level of syntax highlight a double trend towards technicalization and popularization of discourse.

Institutional reports are also the topic of the next chapter, by CRISTINA PENNAROLA and GERMANA D'ACQUISTO. Their study examines the genre of the UN Secretary-General's annual report over the last 60 years in the light of recent studies on genre hybridization (Bhatia 2004; Fairclough 2003, 2006). Their analysis of the discursive and lexico-grammatical features of the documents shows significant changes from the detailed, almost day-by-day narration of the early reports to the 'grand narrative' of the more recent ones, which can be regarded as a tribute to the principles and values of the United Nations. The different use of narrative structure, viewpoint and modality over the years points to an evolution in the very function of the document, which has turned from an informative account addressing the Member States' representatives and UN staff into a persuasive and self-promotional genre, meant for a general readership. Such evolution can be seen as a reflection of major sociopolitical changes, and in particular of the impact of the Internet which has greatly contributed to the democratization of political decision-making and also stressed the need for greater accountability in all kinds of organizations.

Insofar as institutions are public-oriented organisations, communicating with their key publics is crucially important for them, especially when their decisions have a direct impact on the everyday lives of citizens. Within the EU, these decisions are typically entrusted to genres – most notably the legal genre of the EU Directive – which are often too specialised to be understood by the general public. VANDA POLESE and STEFANIA D'AVANZO's chapter investigates EU 'Food safety' discourse in EU Summaries as a form of popularisation of the specialised genre of the Directive. Their diachronic comparative analysis of these two genres is based on Summaries of EU Directives within the thematic area of 'Food safety' in a time-span covering 1982 to 2008, and focuses on the potential ideological implications of the type of discourse being operationalised in the Summary as a popularised genre. Besides showing a growing concern on the part of the EU for food safety problems, peaking in the 1990s, the investigation highlights the role and function of forms of 'citation' as well as the role and value of the modal verb *shall* – typical of legal genres, but not necessarily expected in texts designed to report on them – in the popularising genre of the summaries. The authors conclude that the discursive strategies identified in the corpus are functional to reinforcing legitimation and credibility in the EU involvement in food safety in the 1990s, a decade of great concern for safety, good quality and nutritional value of food products.

The discursive construction of food safety, combined with sustainability-related issues, is also at the heart of PAOLA CATENACCIO'S chapter, which marks a shift from institutional to business communication. Her study analyses the evolution of the discursive strategies deployed in Monsanto's Pledge reports (i.e. documents reporting on the US-based biotech multinational's sustainability initiatives) over the ten-year period from 2000 to 2009. Catenaccio's analysis shows that despite resembling in many ways Corporate Social Responsibility reports, Monsanto's Pledge reports have a much broader agenda, which is to build consensus for genetically modified seeds, of which Monsanto is a world-leading producer. The analysis of the strategies enacted by Monsanto to build such consensus over the period considered shows that the reports engage in progressively more extensive dialogically-framed argumentative discourse aimed at countering challenges commonly levelled at biotechnology. Over the years, they build a corpus of arguments which are taken as a given in every successive document, gradually broadening the common ground Monsanto expects to be able to rely on when communicating to the public at large. Thus, having being tackled once, in later documents still controversial issues are posited as no longer being objects of debate, a fact that creates an impression of growing scientific consensus which is exploited by the company to foster the acceptance of biotech products among the general public.

The use over time of apologetic discourse in CEO's and Chairman's Statements in annual company reports (ACRs) and the identification of its role, if any, in genre variation, is the subject broached by CINZIA GIGLIONI in the next chapter. The study compares reports issued respectively in 2000, the year before corporate accounting scandals and September 11, and in 2009, a time of crisis in the markets worldwide, using an analytical model which distinguishes four different strategies in apologetic discourse. Findings show variations in the genre which are largely connected with the difference in economic outlook between the two periods considered, the first one relatively unproblematic and the second one bedeviled by one of the most devastating crises ever. In particular, in hard times recourse to apologetic discourse is more extensive, relying on all four strategies, but it is not absent in periods of economic stability, albeit with significant differences, i.e. making nearly exclusively recourse to 'bolstering', a strategy aimed at enhancing company image by linking it to abstract values. It can thus be concluded that apologetic discourse is part of ACRs' distinctive generic resources, associated as it is with companies' needs to account for their strategic choices and actual performance. Thus, it may be argued that the market crisis has enhanced an existing trend in corporate communication strategies by encouraging companies to select rhetorical tools suitable to deployment in times of economic downturn, thereby acting as a catalyst in the generic evolution of corporate narratives.

The next chapter, by SYLVAIN DIELTJENS and PRISCILLA HEYNDERICKX, is a diachronic study of CEOs' New Year messages based on historical materials collected from the archives of different Belgian companies. The New Year message, a rather stable genre, with a standard structure, a predictable content and a stereotypical language use, is addressed to the employees and traditionally appears in the December or January issue of corporate magazines. Dieltjens and Heyndericx's study not only shows how this genre changed in the last decades of the twentieth century, but also highlights differences between New Year messages written in times of crisis and those written in more stable times. Over the years, although the texts become longer, the amount of attention given to wishes decreases. Also, CEOs feel less obliged to thank their employees for their work and their efforts. On the contrary, they make stronger appeals for more dedication, and the business-related sections of the messages become more elaborate. These aspects are obviously determined by the fact that the more recent messages were issued in times of crisis, a hypothesis which is confirmed by the other characteristics of the messages, e.g. the more abundant details given about the local and worldwide causes of the crisis and about the measures that will be taken to counter its effects.

Dieltjens and Heyndericx's chapter on this well-established, though not widely studied genre, is followed by a chapter authored by PAUL GILLAERTS focusing on another business genre – job announcements – which, despite its ubiquity, is also understudied both synchronically and diachronically. In his chapter, Gillaerts takes a diachronic perspective, describing the evolution of job ads in the Flemish newspaper *De Standaard* over the last sixty years (1950-2010). His analysis goes beyond description, and attempts to offer an explanation for the development of the job ad by bringing together textual findings and contextual evolutions in Flemish society and corporate communication. Gillaerts's findings suggest that despite its diachronic development, the genre shows an impressive stability in its moves, while at the same time deploying a growing variety of strategies/steps which are indicative of genre bending and mixing with other genres, such as direct mail and commercial advertising. Metadiscourse analysis reveals the growing importance of ethos (stance) in the genre, testified by the increasing use of boosting and attitude markers, and pathos (interaction), shown by the increased use of engagement markers. Both are indicative of the marketization of the genre. Content analysis reveals the emergence of new themes, such as extralegal forms of remuneration and ecological issues, which are immediately linked to the job seekers' interests. Finally, regarding the context of the genre, while the explosion of related (very often digital) genres within the genre system/network of recruitment marketing has pushed the traditional printed job ad to the periphery, the job ad has managed to survive as a 'metagenre', giving birth to ever new genres that enter the centre of the genre system.

With FRANCESCA SANTULLI's chapter, the volume draws to a close. Indeed, the last chapter ideally closes the circle opened by Giuliana Garzone's introductory chapter, revisiting from a different perspective the issues discussed in Garzone's essay. Santulli's starting point is that in order to understand genre it is necessary to consider both its synchronic manifestations and the constant evolution of its individual forms. The functioning of genres is impossible to explain without considering how (and possibly why) they inevitably change: if on the one hand single textual instances can be better understood against the background of shared generic norms, on the other hand it is necessary to bear in mind that they occur in a diachronic dimension and thus contribute to the evolution of the very norms which generated them. From this point of view – argues Santulli – it is easy to find similarities between language and genre evolution, thus making it possible to apply to the diachronic analysis of genres some general principles that have emerged in the long – and fruitful – tradition of studies on language change. Santulli then goes on to illustrate her theoretical approach with reference to the evolution of the tourist guide. Her findings confirm that the historical context plays a crucial role in genre development and evolution, as it gives rise to specific needs, typical of a certain social pattern, which combine with ideological objectives (in the case of the tourist guide, with the aim to offer an evaluative description of tourist sites and at the same time define the identity of modern tourists). They highlight how a new genre can emerge for historical reasons through a re-interpretation and combination of features typical of other existing genres, displaying, at the same time, special distinctive traits.

Overall the studies collected in this volume testify to the extensiveness and rapidity of the change that domain-specific genres have undergone in the last few decades. If one considers the socially situated nature of genres, this pervasiveness of change comes as no surprise in a historical period characterised by a continuous state of flux at all levels – social, technological, economic, philosophical, ethical, spiritual. As the various case studies discussed in the book confirm, genres are sensitive to a whole range of diversified variables capable of inducing change: in this respect, particularly important in the last few decades have been the increasing orientation towards marketization and promotion, and the impact of advances in communications and in computer-mediated technologies. At the same time, in spite of their dynamism and fluidity, genres have actually survived thanks to the inherent persistence in each of them of an invariant core of distinctive features that have contributed to preserving their identity – at least to some degree – even across the most pervasive waves of change.

In this context, the genre analyst's task is the production of an accurate contextually situated and explanatory description, and at the same time the continuous updating of such a description.

But the real challenge is to refine the analytical tools used in the analysis of ever changing genres, adapting them to newly emerging exigencies and tailoring them to novel genres that may arise in the process and which may require new conceptualizations. The studies in this volume are a first step in this direction, and provide preliminary insights into new, promising applications of genre theory to evolving and emerging genres.

Giuliana Garzone

Why Do Genres Change?*

1. Introduction

In a world that is increasingly interconnected and globalized, information and communication flows have intensified more than one might have predicted only few decades ago. By necessity this has led to a rapid diversification and evolution in the texts, genres and discourses exchanged: as Bakhtin points out, "each sphere of activity contains an entire repertoire of speech genres that differentiate and grow as the particular sphere develops and becomes more complex" (Bakhtin 1986: 60). This phenomenon of 'generification' – as John Swales (2004: 4-6) has proposed to call it – involves an ever more rapid process of change, with the evolution of existing genres, their diversification into a variety of sub-genres, and the emergence of totally new ones.

1.1. Aim and method

Generification in domain-specific and professional communication is the object of investigation in this chapter, which looks at the extremely rapid process of genre change that has occurred in the last few decades, and aims to identify and systematize the main factors that have contributed to determining it. Here the expression 'genre change' is used in broad terms, to indicate not only the process of evolution undergone by many existing genres, but also the appearance of new

* The research on which this chapter is based contributes to the National Research Programme *Tension and Change in Domain-specific Genres* funded by the Italian Ministry of University (COFIN Grant No. 2007JCY9Y9).

genres, especially – but not exclusively – under the pressure exerted by globalization and the spread of new technologies.

The basic methodological framework of this study is in discourse analysis, and in particular in genre studies. Today the three main traditions in this area of research that Hyon identified in the late 1990s (Hyon 1996) – the North American tradition (e.g. Miller 1984/1994; Freedman/Medway 1994; Berkenkotter/Huckin 1995; Bazermann 1988, 1994; Devitt 1991, 2004) associated with New Rhetoric, the SFL-oriented Australian tradition (e.g. Halliday/Martin 1993; Martin 1993; Paltridge 1996, 1999, 2001), and the ESP/EAP tradition (Bhatia 1993, 2004; Swales 1990, 2004) focusing on specialized, professional and academic communication – have partially converged and intertwined, if only because in time prominent scholars in each tradition have moved and worked in different universities around the world. Today they share an increasing interest in the exploration of the social practices and contextual variables associated with genres (e.g. Bhatia 2004), with some scholars – especially in the ESP tradition – arguing in favour of the integration of linguistic analysis with ethnographic research based on observation and interview for the purpose of gaining insights into the roles and perceptions of practitioners and into the process of discourse production (cf. e.g. Candlin/Maley 1997; Bhatia 2008a, 2008b, 2010; Flowerdew/Wan 2010).

The common premise on which all these studies rest, representing the distinctive element of modern scholarship in this research area, is the exquisitely social nature of the genre construct, viewed as an interface between language, situation and action. In Jim Martin's words, genres are "configurations of meaning that are recurrently phased together to enact social practices" (Martin 2002: 269). This notion, which has been variously re-elaborated in subsequent theorizations, was already present in Carolyn Miller's original definition of genre as a form of "typified rhetorical action based in recurrent situations" (Miller 1984/1994: 31).

Accepting this premise means interpreting genre in pragmatic terms as a form of rhetorical action, a standardized discursive response to similar situations that people have to face recurrently. Generic conventions put at the speaker's disposal the experience of those who have preceded him/her in tackling the same situations in the form of regularities in the cognitive structuring of text that reflect "accumulated and conventionalized social knowledge available to a particular discourse or professional community" (Bhatia 1993: 21). Thus the individual can meet his/her communicative needs and realize his/her own intentions by means of socially shaped textual forms forged in time as the most suitable for an action that is rhetorically effective and recognizable by recipients. As Miller put it: "A genre is a rhetorical means for mediating private intentions and social exigence" (Miller 1984/1994: 37).

As regards specialized communication in particular, genres are part of professional practices and the acquisition of an adequate command of them is an integral part of the basic process of socialization of anyone aspiring to become part of a professional community (Bazerman 1988): as Paré (2002: 69) points out, "learning to participate in workplace genres means learning one's professional location in the power relations of institutional life".

Against this background, this chapter aims to identify the main factors involved in genre change in the contemporary world, looking at how variables associated with the various participants and contextual factors interact and consequently cause genres to change or diversify in time, or determine the rise of new ones.

1.2. Genre and change

In the by now extensive literature on genre in non-literary domains, research has often focused on the inherent dynamism of genres, without however paying specific attention to the mechanisms involved in evolution. For instance, when Fairclough (1993) identifies an increasing tendency to marketization of discursive practices in British universities' communicative efforts, he does analyse an aspect of the recent evolution of various genres within the academia, but his focus is on marketizaton or 'commodification' more than on genre change. More closely focused on the distinctive characteristics of genre are some of Bhatia's works on 'genre bending', highlighting the promotional 'bend' taken by many business genres in recent years (Bhatia 2004), as well as his studies of genre mixing and genre embedding (Bhatia

1994, 1995), two tendencies that he interprets as evidence of the dynamic complexity of genres, and of their "power to innovate" (Bhatia 1997b: 363).

A work that tackles squarely the problem of genre evolution and the factors involved in it is Berkenkotter's (2008a) study of the psychiatric case record, which she uses as a case study to investigate the "relationship between sociohistorical changes occurring within psychiatry as a clinical profession and changes to the case history's form and content" (Berkenkotter 2008a: 178). She rejects an evolutionary model of genre change imported from the biological sciences as proposed by D.H. Hull and adapted by Gross *et al.* (2002), arguing in favour of a Kuhnian model based on paradigm shifts or 'revolutions' (Berkenkotter 2008a: 189), and concludes that "the genre analyst needs a repertoire of techniques for capturing the complex interaction between socio-historical, technological, demographic, and epistemological factors in order to track and interpret textual changes in disciplinary texts" (Berkenkotter 2008a: 189; cf. also Berkenkotter 2008b). This is in line with the insistence on the attention to contextual and social elements in approaching genres that prevails in contemporary genre studies.

Here genres will be investigated in a short-term diachronic perspective with the purpose of identifying the main forces at play in driving change. The underlying assumption is that, in spite of the enormous complexity of genre evolution in specialized and professional communication, it is possible to identify some general patterns, and conceptualize them in different typologies on the basis of the factors inducing change.

2. Factors inducing genre change: changes in professional and disciplinary practices

In this context, the view of genre as a social artifact responding to a whole range of external and pragmatic variables is the inevitable starting point if one is to shed light on the main patterns characterizing the evolution of genres over time, as it implies that change occurs in response to modifications and developments in the social, cultural and institutional contexts genres are imbricated in (cf. Kress 1987: 42; Devitt 2008: 89-91).

The first change-inducing factor to be examined here regards evolutions in professional and disciplinary practices, which is the result of the close relationship between genre and *local* social practices. This relationship has been proved and widely explored in a rich and diversified literature. Within the general picture, there are especially prominent scholars, like Vijay K. Bhatia, Charles Bazerman and John Swales, whose contributions – published from the late 1980s to this date – have been crucial for the accumulation of current knowledge in genre research. In the last three decades their works have provided important frames of reference for researchers, offering fundamental theoretical insights and dealing with a wide range of genres, problems and issues in various scientific and professional domains.

More in general, there has been a multitude of scholars conducting research on genres in professional, disciplinary and scientific domains, highlighting the relationship between domain specific social and discursive practices and genre use. Many of such works fall within the category that Bazerman, in a recent methodological article, has defined "theories of the middle range" (2008: 302), as they keep account of broader social and cultural aspects, and at the same time examine particular textual and generic phenomena, thus combining theoretical, empirical, and practical concerns. The examples are numberless; just to mention a few: in disciplinary discourse, Atkinson (1999) and Gross *et al.* (2002) on scientific genres, Battalio's (1998) on ornithological discourse; in professional discourse Gergen/Graumann's (2006) on genres in psychology, Berkenkotter's (2008a) work on psychiatric case history report already mentioned above, Yates' (1989, 2005) on business correspondence and on insurance, Campbell/Hall Jamieson's (1990) on political genres, and in particular on US presidential addresses, and also all the articles collected in Barton/Hall's (1999) and Gillaerts/Gotti's (2005) edited volumes on the letter genre.

Whether taking a synchronic or (less frequently) a historical approach, the findings of these works confirm that in domain-specific communication, genre is strictly dependent on, and contributes to the shaping of, social practices, and therefore it is sensitive to any minute variation in them, changing accordingly.

2.1. Interdiscursivity

Among such variations, one that has attracted substantial attention in research in the last few years is interdiscursivity (Candlin/Maley 1997; Bhatia 2010), which Bhatia defines as the "appropriation of semiotic resources (which may include textual, semantic, socio-pragmatic, generic, and professional) across any two or more of these different levels, especially those of genre, professional practice and professional culture. Appropriations across texts thus give rise to intertextual relations, whereas appropriations across professional genres, practices and cultures constitute interdiscursive relations" (Bhatia 2010: 35). This notion has a great heuristic value to account for genre change today, and in particular for changes that are brought about by developments in the practices of specialized discourse communities deriving from contacts with other discourse communities that determine forms of contamination and overspills in discursive practices and communicative conventions.

An exemplary case is the contamination of arbitration discourse practices with those typical of court proceedings, a phenomenon, which has received extensive attention from sociolinguists and discourse analysts in the last few years (cf. e.g. the studies collected in Bhatia/Candlin/Gotti 2004 and Bhatia/Candlin/Gotti 2010). Bhatia (2010: 37) uses it as an exemplary instance of interdiscursivity, i.e. the appropriation of alien resources in terms of practices, which he sees as a case of "colonization of arbitration by litigation" (Bhatia 2008b: 169-170). While in the last few decades arbitration has become ever more popular for the settlement of commercial disputes (Hunter/Marriot/ Veder 1995; Mattli 2001) as an alternative to court proceedings in consideration of its confidentiality, its relative rapidity and its total autonomy with respect to all forms of local judicial procedures (i.e. national courts), today international arbitration is subject to an increasing influence on the part of litigative procedures. This is due to the ever more frequent tendency to employ lawyers as arbitrators, bringing with them their legal background and expertise, and the increasing institutionalisation of procedures (Bernardini 2000; Miles 2003; Nabatchi 2007) with the passing of specific arbitration legislation in many countries. Hence the progressive 'colonization' and 'judicialization' (Nariman 2000: 262; Marriott 2004: 354) of arbitration in terms of techniques, procedures, shared expertise and discursive practices, but also an increasing tendency of text genres produced within the context of arbitration institutions to be contaminated in terms of text organization, drafting style and even argumentative preferences typical of the relevant legal and judicial systems (cf. e.g. Garzone 2004a and 2010, respectively on arbitration rules and on arbitral institutions' website communication; Maci 2010 and Degano 2010 on awards).

2.2. Imposition of conventions

If evolutions in professional practices and cultures have been an important driving force in genre change in domain-specific communication, in some cases change has been induced or accelerated by deliberate imposition by institutions or other 'powerful' subjects.

A case in point is that of the scientific research article, which in recent times has undergone further standardization as a genre, with a high degree of compliance with the prevailing conventions in terms not only of language and style, but also of possible contributions and content organisation. This is not the result of a spontaneous evolution or convention consolidation, but rather the effect of deliberate imposition on the part of the most important international journals. They have acted as repositories of generic conventions and have formalized the discursive structures that are deemed to be most suitable to record the operational and argumentative procedures applied in research within the scientific paradigm associated to each discipline. In this case, the gatekeeping action of international journals has imposed changes that were inherently desirable, embodying the paradigm prevailing in the relevant discourse communities (Bazermann 1988), but which if unguided would have taken a much longer time to occur.

In the professional domains, something similar can be said about legislative drafting conventions. A change in the complexity and verbosity of legislative texts has been advocated by Plain Language campaigns and other organizations for decades now (cf. Cutts 1999, 2000; Locke 2004). In some cases this has triggered actions by governments openly imposing changes in writing conventions in the legal and administrative field (cf. e.g. in the U.S. Clinton's *Memorandum for the Heads of Executive Departments and Agencies* 1998; in Italy Circolare del Presidente del Senato 20 aprile 2001 *Regole e raccomandazioni per la formulazione tecnica dei testi legislativi* [*Rules and recommendations for the technical drafting of legislative texts*]). But even where this has not occurred, and in particular I am referring here to the United Kingdom, the awareness-raising action of the various campaigns and of the flurry of works that have been published in the last two decades advocating simplification in legislative drafting have resulted in a deep change in the structure and wording of legislative texts, especially since the turn of the millennium, with long paragraphs broken down graphically into separate clauses/sections and a modified use of modals (cf. Garzone 2008).

While in the foregoing Sections (§2, §2.1 and §2.2) the discussion concentrates on the impact of evolutions in professional and disciplinary practices on genre change and some of the mechanisms at play in this interaction, in the next Section attention will shift to genre change brought about by evolutions affecting society at large.

3. Genre change induced by broad societal changes

Although the relationship between discursive practices and genre in professional/disciplinary domains has been widely explored in the literature, albeit not often with a specific focus on change, less attention has been given to the impact of broad cultural and societal changes on genres in non-literary communication (e.g. academic, professional, political, organisational, legal, journalistic, etc.). This impact – I want to argue here – is really extensive, in spite of the relatively limited interest it has prompted until recently.

3.1. The Corporate Social Responsibility Report

In the contemporary world, some meaningful examples can be found in developments in external corporate communication. In a generic perspective, especially interesting is the appearance of Corporate Social Responsibility Reports as a section of Annual Company Reports (ACRs), gradually evolving into a separate companion document (e.g. Kolk 2008). The appearance of CSR Reports responds to a new attitude of the general public towards the corporate world. In turn, this attitude reflects the profound changes in the structure, organization and status of the business enterprise in post-industrial society, which today is constructed as accountable not only to shareholders and customers, but also to a whole range of indirect stakeholders and other social actors.

Although in one form or another CSR communication had already marginally existed since the early decades of the twentieth century (Catenaccio 2011a: 169-170), the CSR Report – and its various realizations as Sustainability Report or Environmental Report – started to be issued regularly in the mid 1980s either as a distinct section within a company's Annual Report or as a separate document, and consolidated its presence starting from the turn of the millenium (cf. e.g. White 1999; Kolk 2008, 2010; Brown *et al.* 2009). According to a recent report (KPMG 2011), 95% of the 250 largest companies worldwide now issue CSR reports, with an increase of more than 14 percent over the same 2008 report. On the other hand, a survey focusing on medium to large privately held businesses worldwide shows that the companies actually reporting on CSR/sustainability are about a quarter of the total globally, although a vast number of businesses across the globe continue to be driven towards more ethical practices and increasingly take some form of CSR initiative (cf. Grant Thornton 2011: 16).

Originally a section of the ACR – and in some cases still so –, the CSR report is now developing into an autonomous text with a configuration that follows broadly the blueprint of the parent genre, being usually preceded by the CEO's message or letter, parallel to Shareholders' Letters traditionally included in all ACRs. It is in an advanced state of development as an independent genre, also thanks to the adoption of uniform international standards within the framework of the Global Reporting Initiative,[1] and exemplifies very well the emergence of a new genre, or better its coming into existence as an offshoot of a previously existing one, under the pressure of emergent societal values and ideological stances. Its name is still relatively unstable too, with many companies using the general denomination 'Corporate Social Responsibility Report', while others, especially if operating in a high environmental impact sector, prefer 'Sustainability Report'.

In terms of genre content, if ACRs consist of multiple components, such as Letter to Stockholders, company profile, company overview, financial highlights, operational overview, which some authors consider to be autonomous sub-genres (e.g. Jameson 2000; cf. de Groot 2008: 35ff.),[2] the CSR report also tends to be divided into separate sections, although discursively less detached from each other as the main focus is constantly on values and sustainability, including a CEO's message (sometimes called 'letter'), which is more detached from the rest of the Report than all the other sections, and in this respect has the same position as the Shareholders' Letter in Annual Reports. Recent research has shown that overall the generic content of CSRs seems to be relatively uniform, thus confirming a consolidation of the genre is under way, although there certainly is a degree of variability as a function of the sector in which each company operates and its priorities (cf.

1 The Global Reporting Initiative (GRI) is a network-based organization that pioneered the world's most widely used sustainability reporting framework. Its core goals include the mainstreaming of disclosure on environmental, social and governance performance. Its Reporting Framework has its cornerstone in the Sustainability Reporting Guidelines, now in their third version, known as the G3 Guidelines, published in 2006. The Guidelines set out the principles and performance indicators that organizations can use to measure and report their economic, environmental, and social performance. Cf. <http://www.globalreporting.org> (unless otherwise stated, websites were last accessed 10/02/2012).

2 While the status of the various chapters of ACRs is not so clearly one of autonomy, this is the case for the Letter to Shareholders, on whose independence there seem to be convergent opinions and a specific literature (cf. e.g. Nickerson/de Groot 2005; Garzone 2005a).

e.g. Catenaccio 2011a; Malavasi 2011). As Catenaccio (2010) has shown, in the CEO's message it is possible to identify the three main moves already identified in Letters/Messages to Shareholders (cf. Garzone 2005a: 194ff.), albeit with slight variations (e.g. Move 1, *Reporting on Company's performance, trends and results in the relevant year,* and Move 2, *Providing a narrative of salient facts*, tend to be conflated) and with the addition of an initial move (*Defining CSR / Justifying CSR involvement with a view to changing existing conceptions of the role of business in* society) and a final one (*Soliciting response*) (cf. Catenaccio 2010: 193-197).

Overall, it is possible to say that, today, the CSR report has come into existence as part of the evolution of the ACR, and has developed into an autonomous genre, with its peculiar generic components, and this is the result of the pressure of new ideas and values associated with important societal changes.

3.2. Green advertising

Another example of genre change brought about by the very same developments in society's values and orientations (Catenaccio 2011b) is the gradual spread of advertisements that focus more on companies' environmentally and socially responsible policies than on product promotion, the so-called 'green ads' (e.g. Hansen/Machin 2008). These are ads that use environmental themes in order to attract customers so that they "consistently choose one product over another for environmental reasons (packaging, formulation, advertising)" (McIntosh 1991: 208). The messages deployed insist on the ecofriendly characteristics of the product, on the basis of the premise that to do something for the environment is a priority for the consumer. This is obviously part of an enthymeme that takes this premise for granted, while of course it would not be unreasonable to think that for many consumers other variables are more important (price, quality, convenience, etc.).

At any rate, the consideration that a message focusing on environmental issues should be used to promote products which are genuinely ecofriendly, or at least potentially so (foods, detergents, cosmetics, packaging, etc.) is rational and predictable. But it is more interesting that it appears also in the ads of companies whose activities and products are by definition polluting, like motor vehicles and oil-derivatives.

Green advertising for cars is inherently fraudulent because – however fuel-efficient and low-emission they could be – they are always polluting, as is proved by the fact that Norway put a ban on green car ads in 2007.[3] Similarly environmentally adverse is the impact of a whole range of other industries, first and foremost the oil industry, where most companies – above all those involved in judiciary cases and environmental accidents – launch 'green' campaigns to bring home a renewed environmental reputation.[4]

In actual fact, in society the attention for environmental problems is so strong – albeit not necessarily serious and based on a real awareness of the issues involved – that many companies have made the choice to rely on these themes for successful advertising and promotional campaigns, independently of their real policies. In many cases this card is played even for non-environmentally relevant products, such as clothing (cf. e.g. Diesel jean's 'Global Warming Ready' campaign).[5] As already in 2007 an article in the *Wall Street Journal*[6] pointed out: "Green is the new black. Madison Avenue tried to curry favor with consumers this year by coloring products and brands with an environmental tint. A long list of companies such as General Electric Co., Chevron Corp., and Home Depot Inc. all jumped on the eco-friendly bandwagon."

And it is true that, in many cases policy statements are not accompanied by a genuine intention to implement them (cf. Ramus/ Montiel 2005: 409). Green advertising thus becomes 'greenwashing', i.e. "the misuse of the principles of environmental marketing" and means that consumers cannot trust the content of advertisements (Kärnä *et al.* 2001: 59).

---

3 Cf. <http://www.terrapass.com/blog/posts/norway-outlaws>.

4 For a discussion of some examples of the strategies used by companies in this type of campaigns, cf. Catenaccio (2011b: 363-368).

5 Cf. <http://theinspirationroom.com/daily/2007/diesel-global-warming-ready>.

6 "Best of Ads, Worst of Ads". 27/12/2007, B1.

Whatever the relevance of the notion of sustainability to the product being promoted and the degree of honesty of companies' 'green' statements, it is a fact that a profound change in society's values and perceptions has had a deep impact on a well established communicative genre, advertisements, determining the emergence of new sub-genres characterized both by new distinctive discursive patterns deriving from the commercial appropriation of the discourses of global warming and climate change for promotional purposes, and by the occurrence of new generic components, environmentally oriented, to achieve the same purpose as before, i.e. promote sales and enhance corporate reputation.

4. Technology-driven change[7]

In the cases discussed so far, genre change – whether spontaneous or imposed – is strictly linked with social or, in more general terms, cultural factors that have emerged in the primary or secondary culture of a given discourse community, or in society at large.

But the most profound changes have occurred as a result of the spread of new technologies. They have been induced by the unprecedented options offered by new media and communicative environments, mostly associated with Computer- and Web-Mediated Communication, but also with communications technologies, and in particular mobile phones, whose affordances are increasingly converging with those made available by computers (Twitter is a case in point).

These innovations have opened up new, unexpected and virtually boundless discursive, as well as social, spaces. New media have brought with them unprecedented ways of communicating and new modes of discourse, radically changing the semiotic configuration of communicative events, and revolutionizing our most basic notions about communication. Since – as Yates and Orlikowski point out – "medium may play a role in both the recurrent situation and the form

7 This section is loosely based on Garzone (2007).

of a genre" (1992: 310), the spread of new computer- and Web-based technologies has inevitably had a deep impact on genres, first of all in terms of mode of discourse, giving rise to new social practices, and radically altering the semiotic coordinates of communicative events. Hence very extensive genre change. For many genres now there exists a multimodal 'double' resulting from their migration to the web, while many new genres have been born that are native to the web, and have unprecedented peculiarities in terms of mode of discourse (e.g. Askehave/Ellerup Nielsen 2004; Garzone 2007).

In order to trace changes in existing genres fostered by the spread of Web Mediated Communication, the incorporation of the notion of medium into traditional genre analysis – as advocated by Askehave and Ellerup Nielsen (2004, 2005) – is not sufficient, because the Hypermedia Computer-Mediated Environment (HCME: cf. Hoffman/Novak 1996) is not simply one medium, but rather an environment providing access to various media, and is characterized by a whole range of distinctive properties that have affected genre evolution.

4.1. Multimodality

The first distinctive discursive property of the HCME is the availability of a full range of multimodal affordances, i.e. the possibility it offers to deploy different semiotic resources within a single communicative process. Although in theory multimodality is an inherent property of human communication, characterising most of its forms (think of the combination of audio and visual input in ordinary face-to-face interaction; cf. Stöckl 2004: 10; Santulli 2007: 31-33), it is made readily available in the computer-mediated environment, which gives simultaneous access to the use of a number of multimodal technologies, images, sounds, animation, films, which can be easily combined and utilized together with no special technological procedures nor additional costs for reproduction and distribution. Thus, recourse to complex graphics, intensive use of pictures and photographs, animated features and special effects is a common trait of web-genres. Incidentally, it is interesting to note that to some extent this fact affects also online texts in 'frozen' formats, which are subject to a sort of 're-mediation' effect

(Bolter/Grusin 1999). This is evident if one examines documents that are made available in both HTML and PDF format on the Internet, as is often the case, for instance, of annual company reports, whose PDF version usually contains a reconstruction of the hypertext in the form of a series of pages in a sequence, with no changes in content or form, and exhibits the same impressive range of graphs, charts and vivid colour images as the HTML.[8]

The inherent multimodality of the HCME makes it necessary for the analyst, when investigating Web discourse, to adapt pre-existing language-centred analytical frameworks in order to accommodate a multiple semiotic makeup (Garzone 2002).

4.2. Extension in participation framework

Another important feature of online communication, having an important impact on genres, is the potential global reach provided by the WWW, i.e. access to a global audience. This is what Kress and van Leeuwen call "distribution" (2001: 7-8; 103-110), i.e. the possibility offered by web-mediated communication to get in touch with millions of users the world over instantaneously.

Drafters of documents to be published on the Web are aware that their potential audience does not comprise only their primary addressees, but also a virtually infinite number of Internet surfers who may simply come across the document by chance and then become readers and, hopefully, users, followers, customers, investors, etc. In textual terms these participants have the status of ratified participants rather than simple bystanders (Goffman 1981: 131-133), and as such they are constructed in the relevant texts. The enlargement of the potential audience has also contributed to 'genre bending' (cf. §1.2 above), and in particular to adding a promotional component to most genres that have migrated to the Web (e.g. Annual Company Reports and CEOs' Letters; cf. also Garzone 2004b).

---

8 Cf. for instance the analysis of Campbell Soup Annual Company Reports (<http://www.shareholder.com/campbell/reports.cfm>) in Garzone (2009).

At the same time, Web-mediated communication offers an opportunity for individualized contact and expression, as on the one hand it addresses each user individually and, on the other, gives everyone the possibility of exchanging views with other users – either synchronously or asynchronously. This point will be discussed shortly when dealing with the role of users in web-mediated communication.

4.3. Hypertextuality/hypermediality

The property of the HCME that has had the deepest impact on genre change is hypertextuality, which is the constitutive characteristic of Web text. A hypertext, according to Theodor H. Nelson, "is a series of text chunks connected by links which offer the reader different pathways" (Nelson 1981: 2). When created for a given purpose and stored on a CD or any other magnetic support, hypertexts – albeit vast and complex – are closed systems. But the WWW in its global interconnexity represents the conflation of an immense number of hypertextual systems, thus realizing – as J. Landow (1992: 3-4) points out – Roland Barthes' description of ideal textuality (Barthes 1970/1974): a galaxy of signifiers, with no beginning, reversible, accessible by several entrances, made up of systems of meaning the number of which is never closed, based as it is on the infinity of language (Barthes 1970/1974: 5-6 [English translation]; 11-12 [French]).

Hypertextuality has a number of consequences that deeply condition texts, discourses and genres. A first, superficial (but not for that unimportant) consequence is that when used to construct a hypertext, text is to be broken down into sections each of which will constitute a node, hosting objects as well as hyperlinks leading to other nodes. This is what is called the 'grain' of a hypertext, which mostly corresponds to the amount of text that can be displayed on the screen, but it can be smaller, as in the hypertextual environment text is mostly presented either in the form of *lexias*,[9] i.e. graphically separated blocks of

9 The word *lexia* was introduced into English by Landow (1992: 34) to translate Roland Barthes' term *lexie* (1970/1974), the meaning of which broadly corresponds to that of Derrida's *morceau* (1972).

text, often coinciding with paragraphs, limited in size and often in width, or divided into columns, in the style recommended by Web-writing handbooks. Usually this regards mainly the pages on the surface levels of Web sites, which are more genuinely hypertextual, while the deeper levels often contain lengthy stretches of text, obviously destined to be printed for efficient reading. In the case of longer documents, the software used to encode and post the text is fundamental: PDF files are essentially frozen and present the text in a continuous whole made up of the 'picture' of the sequence of printed pages; when html is used, the text is usually broken down into sections, each accessible from a main menu, and scrolling is adopted when a section is longer than the amount of text contained on a screen.

A good example is that of the texts of English Acts of Parliament. The original printed text is published in a booklet with an initial table of contents followed by a continuous text, a few pages long. The electronic version available on the Website of the Office of Public Sector Information[10] has a menu on the first page, and each item on the menu is a hyperlink that directs to a page containing the corresponding section of the Statute. Thus Web-mediated communication can present "meaning-by-degree" (Lemke 2006): the text is not made accessible to the reader all at once; rather, it unfolds in discrete units to which access is given by means of navigation devices. Therefore, each textual unit, each lexia must be decoded and interpreted in relation to the text units to which it is linked within the overall textual system (Mitra/Cohen 1999).

Another important effect of the hypertextual environment is that it involves the activation of different reading modes. Alongside the traditional linear 'reading-as-such' modality, hypertextual environments also involve a non-linear reading modality, i.e. hyper-reading (Sosnoski 1999: 135): by switching to the 'navigating mode' (or 'linking mode') the reader can navigate the site and actively construct his/her own reading path through one or several sites (Askehave/Ellerup Nielsen 2004, 2005).

From a generic perspective, it is interesting that (hyper)links, the main functional elements that enable navigation, realize moves in

10 URL <http://www.opsi.gov.uk/acts.htm>.

a way that does not exist in the reading of printed texts. Not only does the addition of links to a Web page enable the navigator to 'move on', but has also the function of connecting various items on the web page (and also outside it), setting up a semantic relation between them, and with other webpages. In this way, designers conceive the architecture of a hypertext in the form of a number of possible navigation options, prefiguring a range of pre-established inner navigation paths and external links leading to other sites.

In this context, the user is in a position to materially contribute to the realization of the hypertext every time s/he explores it (cf. e.g. Garzone 1997; 2002: 286-287), with largely unpredictable results. This obviously interferes with the concept of move sequence and structure; in particular, the variability of discourse structure in text realization afforded by the hypertext medium represents a crucial element to be reckoned with when dealing with issues involved in genre recognition and classification.

All these considerations apply in full to the genre of the website, which was born only few decades ago with the rise of Web communication. It exhibits all the features described above and is the prototypical example of a genre originating from the affordances available within the web-mediated environment, while other forms of web-mediated communication (e.g. email, chatrooms, newsgroups, etc.) tend to share only some of these characteristics: for instance, email messages tend to be nearly exclusively in writing, and hardly exploit their potential hypertextuality, as only occasionally do they include links and are accessed in the traditional reading mode.

Overall, all the features of web-mediated genres described so far contribute to enhancing the role of the user as co-author or, at least, co-articulator of discourse. In some cases, the user is also given the power to add text (or even images, sounds, animations etc.) and create links and pathways, thus "[he/she] more fully collaborate[s] in producing an artifact that is materially different because of [his/her] participation, and not just semiotically reinterpreted" (Lemke 2006: 318). As Lemke (2006) points out, co-articulation is a feature which contributes to avoiding monologism in communication, in the Bakhtinian sense of including multiple social voices (Bakhtin 1935/1981).

In more recent web applications, usually classified under the label 'Web 2.0', this is even more marked, as the user's role involves not only co-articulation, but even collaborative creation. Web 2.0, also called the 'Social Web', comprises a second generation of Web applications that favour dynamic user participation (cf. O'Reilly 2005; Harrison/Barthel 2009). Thus, users do not have a role that is purely passive/receptive or marginally interactive, but rather one of active participation and creation, thanks to 'low' technology that enables users to generate and upload content even in the absence of advanced technical skills. Among such new genres there are discussion lists, chat lines, participatory news articles, weblogs, wikis, folksonomies,[11] and genres associated with social networks, e.g. the Facebook or Myspace profile (Kaplan/Haenlein 2010) and the tweet (i.e. the very short message – less than 140 characters – used for communication on Twitter) (cf. e.g. Marwick/Boyd 2010), which either have an independent, autonomous status, or are sub-genres that contribute to changing the configuration of their 'parent' genres (e.g. weblogs attached to corporate websites or online newspapers).

In light of the foregoing discussion, it can be stated that the analysis of the inherent properties of the discursive and social spaces opened up by web-mediated technologies – multimodality, extension of participation framework, granularity, hypertextuality/hypermediality, co-articulation, interactivity – provides evidence of the pervasive impact that new technologies have had on the generic landscape of communication in the contemporary world, requiring a revision of the notions and criteria traditionally relied on in genre analysis.

---

11 A *wiki* is a web page whose content can be modified, deleted or added to by users via a web browser by means of a 'wiki software'. Wikis are often created collaboratively (cf. Wikipedia, *ad vocem*). A *folksonomy* (a portmanteau of *folk* and *taxonomy* coined by Thomas Vander Wal) is a system of classification derived from the practice and method of collaboratively creating and managing tags to annotate and categorize content. In a social distributed environment, sharing one's own tags makes for innovative ways to map meaning and let relationships naturally emerge. Cf. Wikipedia (*ad vocem*); Quintarelli (2005).

5. Concluding remarks

The picture that emerges from this discussion is remarkably complex and confirms the extreme sensitivity of genre to a whole range of contextual coordinates, social and cultural factors, intertextual/interdiscursive interferences, ideological forces, external pressures, and technological change. All these factors have the potential to induce diversified, or even totally new, rhetorical needs requiring new discursive devices to fulfil them and involving a greater or lesser degree of sociocognitive adaptation.

In particular, what all the factors discussed in this work have in common is that they have some kind of more or less direct social relevance, and thus they inevitably have an impact on genre, which is now universally recognized to be a social artifact responding to a whole range of pragmatic variables. Genre change is triggered by shifts in values and practices originating spontaneously within discourse communities, as well as by values and ideologies pertaining to society at large. Even the innovative effect of the use of new technologies makes itself felt through social repercussions, affecting communicative events' participation framework, altering actors' roles, and making new social spaces available for interaction. The other main category of relevant factors – especially momentous in the case of new technologies and web-mediated communication – are of an essentially formal nature, and are connected with the affordances and/or constraints associated with each communicative environment. Of course, all these factors have to be considered with due caution, without losing sight in each specific case of all the relevant variables, and of the inevitable overlaps and interferences among different factors and agents, and keeping in mind that an effort of this type by necessity involves oversimplification for the sake of conceptualization. This is so in all forms of broad generalizations, but identifying patterns and lines of development is a heuristic endeavour leading to findings that can be used as a starting point in future research which, by enlarging the sample of cases examined and delving more in depth into the possible impact of the different factors involved, may produce reliable and theoretically sound generalizations.

Academic and Scientific Discourse

MARINA BONDI / SILVIA CAVALIERI

The Evolution of the Abstract as a Genre: 1988-2008. The Case of Applied Linguistics

1. Introduction

Academic discourse has been the object of increasing attention in the last two decades, especially from a genre perspective (Swales 1990, 2004; Bhatia 1993, 2004). Interest has focused on empirical research-based genres (research articles, abstracts, etc.) often combining linguistic and rhetorical analysis (Berkenkotter/Huckin 1995). Although not as widely researched as the research article itself, the abstract has recently drawn the attention of a number of genre researchers. As Bondi (1997: 396) states, "abstracts would seem to provide excellent material for genre analysis. Their textual structure is comparatively easy to identify and their size is manageable for different types of linguistic analysis", but particularly so for close analysis of rhetorical moves (Bhatia 1993; Kaplan *et al.* 1994; Dos Santos 1996; Hyland 2000; Bondi 1997, 2001; Stotesbury 2003; Martìn-Martìn 2003, Dahl 2004, 2009, Lorés-Sanz 2004, 2006, 2008, Pho 2008; Gillaerts/van De Velde 2010) or sentence relations (Bondi 2004; van Bonn/Swales 2007; Golebiowski 2009).

Debate around the textual structure of abstracts shows substantial agreement on the role played by the empirical research process. Following Bhatia (1993: 78-79), four moves can be identified: 1. Introducing purpose: a statement of the author's intention (the problem to be tackled, the objectives of the paper, the hypothesis on which the research is based, the thesis supported); 2. Describing methodology: an exposition of the experimental design (data, procedures, methods, scope); 3. Summarising results: a statement of the author's findings and observations on the suggested solutions to the problems; 4.

Presenting conclusions: the interpretation of the results, a statement of the inferences that can be drawn from the findings and the implications and applications they suggest. Some variations to this move structure have also been identified and, for example, Dos Santos (1996) proposes a separate Background Move ('Situating research') before Move 1: 'Introducing purpose'.

A key issue in defining the structure of the genre seems to be the relationship between the abstract and the abstracted article, i.e. the way the structure and content of the article itself are represented, in a multiple semiotic process in which words are used to represent and interpret verbal objects. The relationship between an abstract and the article itself has been variously described in terms of 'news value' (Berkenkotter/Huckin 1995) and of 'representation' (Bazerman 1988, 1994, Bondi 1997). Bondi (1997) looks at the development of the genre in economics discourse and points at an increasing foregrounding of meta-textual and metadiscursive references to the original article and its procedures. Expressions referring to discourse procedures are often given main-clause position, and the types of entities chosen as grammatical subjects are often discourse products and producers (e.g. *the results show, we analyse*) rather than discourse objects (e.g. *cases are discussed, records are analysed*). The issue has great impact on the thematic structure of abstracts (Kong 2004, Lorés-Sanz 2004), but our attention here is rather on the reflexive lexis of discourse representation and its phraseology.

Variation across different types of abstracts has also been noticed. Lorés-Sanz (2004) for example goes back to the distinction between informative and indicative abstracts, often adopted in librarianship. Informative abstracts give the specifics of the research, including results and conclusions. They are characterized by a full IMRD structure, mirroring the organization of the research article. Indicative or descriptive abstracts, on the other hand, do not reproduce the whole development of the paper, but focus on its scope; they deal with what the text is about – issues or problems explored, purpose and methodology, but do not present results. These are rather inspired by the canonical structure of the introduction, described as the 'CARS' model (Create A Research Space) (Swales 1990): establishing a territory, establishing a niche, occupying the niche. A mixed pattern can also be identified. Bondi (2004) looks rather at the differences that can be found in the articles themselves, and are thus reflected in the abstract. Some disciplines and approaches seem to background the process of experimental research and rather foreground the basically argumentative nature of academic discourse: historical abstracts, for example, can be easily divided into those that foreground the narrative of history and those that foreground the underlying argument. As already noted in Bondi (2010: 36),

> [w]hen an argumentational perspective is adopted, the question of whether the abstract is informative or indicative becomes secondary to the need for the abstract to reflect the nature of the argument presented, whether mostly inspired by the typical structure of experimental research (IMRD) or by the argumentative nature of academic discourse, with the identification of a claim and arguments supporting it (and arguments refuting counter-claims).

By guiding the reader's interpretation of the abstracted text, abstracts can be seen to offer a persuasive representation of the argumentative structure of the text and a representation of the writer as a competent researcher. A functional view of the genre thus relates the current relevance of abstracts to the proliferation of scientific production and the need for writers to situate themselves and their work in their disciplines by claiming significance and displaying credibility and membership (Hyland 2000: 63). The writer's identity and position is often constructed in abstracts by displaying membership in a scientific community (for example by emphasizing expertise in the research process) and by making claims about the significance of their own work (for example by emphasizing the argumentative dimension). Self-presentation and writer's identity thus become prominent pragmatic features of the abstract (Lorés-Sanz 2004, 2006).

Keeping in mind the central role of argument in academic discourse and its link with the discourse community, language reflexivity can be studied as a key tool in the representation of discourse. The present study investigates the evolution of the abstract as a genre over a time span of twenty years (1988-2008) considering in particular the field of applied linguistics. The aim is to explore the changes occurring in communicative practices and linguistic patterns. These can be related to the pressure of technological innovations such as databases, increasing the need to compete for the reader's attention, as well as to the spread of the abstract genre across wider fields, leading to writers' greater awareness of the need to write effective abstracts. The study will look more specifically at markers of the 'representational' nature of abstracts, i.e. the foregrounding of metadiscursive references to the original article and its procedures as objects of abstracting discourse, in relation to the type of article abstracted, observing rhetorical and lexical change.

2. Materials and methods

The present study was carried out on two comparable corpora of abstracts, 70 for 1988 and 70 for 2008. They were collected from seven refereed applied linguistics journals (i.e. *Applied Linguistics, English for Specific Purposes, International Review of Applied Linguistics, Journal of Pragmatics, Linguistics & Education, System* and *TESOL Quarterly*), which can be considered representative of a wide range of subdisciplines.

Table 1 gives an overview of the list of the journals selected with their date of first publication and first presence of the abstract. The table shows the rise of the abstract as a genre in the academic discourse community in the last twenty years, following the trends noticed in other social sciences (see Bondi 1997 for economics).

Journal	*First published in*	*Abstracts from*
Applied Linguistics	1980	1987
Iral	1963	1963
Tesol Quarterly	1968	1978
Journal of Pragmatics	1977	1977
Linguistics and Education	1987	1987
System	1973	1973
Esp	(1980) 1986	1980/1-1986

Table 1. Date of first publication and presence of the abstract.

The growing importance of the abstract is also confirmed by its growing length over the last two decades as demonstrated in Table 2:

	Tokens	*Sentences*	*Paragraphs*
1988	10,311	10,579	89
2008	13,202	13,378	70

Table 2. Abstracts' length 1988-2008.

From the quantitative overview of the two corpora, it is possible to observe that the average length of abstracts has increased by almost 3,000 tokens. The number of sentences also increased from 1988 to 2008, whereas the number of paragraphs was reduced to a standardized coincidence between number of paragraphs and number of abstracts. The increase in length seems to confirm what has already been noticed by other scholars (Hyland 2000: 81; Gillaerts/van de Velde 2010: 132).

The methodology adopted for this study proposes an integration of corpus and discourse perspectives. The aim is to explore changes in: a) the generic structure of the abstract, b) the general style used, with reference to personalization and forms of "self-mention" (Hyland 2005) and c) the patterns adopted in "framework sequences" (Bondi 2010), metadiscursive expressions combining forms of self-mentions and frame markers (Hyland 2005) used to represent the structure of the abstracted article.

First, a qualitative analysis was carried out on the two corpora. On this basis we identified two basic types of abstracts: a) *argumentative*, i.e. abstracts in which the focus is on making a claim and defining an issue in the disciplinary context; b) *empirical,* i.e. abstracts in which the main aim is reporting on the research carried out. Abstract (1) is a good example of *argumentative abstract*, whereas Abstract (2) provides valuable evidence for an *empirical abstract*.

(1) *Some initial Considerations for Course Design*

> This paper attempts to outline in theory what a more holistic approach to the design of an ESP course might have to take into account and how a more adequate descriptive base might be built. It begins with a theoretical justification for the inclusion of an ethnographic perspective and then goes on to illustrate how considerations deriving from this and other perspectives

influenced the design of a communication skills course for computer practitioners in the Sultanate of Oman. (ESP 2008-2)

(2) *Disagreeing and Requesting by Americans and Venezuelans*

This paper examines the stylistic devices used by American (L1) and Venezuelan (L2) speakers in two different English language role-play situations: disagreeing and requesting. Although each of the participants used a different combination of devices, there were patterns followed by L1s and L2s that reflected their distinctive cultural styles. In general L1s preferred nonconfrontational stylistic devices when they disagreed with an L1 interlocutor and impersonal stylistic devices when they requested a service. L2s, on the other hand, used more confrontational devices when disagreeing and more personal devices when requesting a service. This suggests that L1s and L2s participated within different frames, respectively 'This is Business' and 'This is (Un)Friendly'. In general these different frames of participation led to more harmonious outcomes for L1s than for L2s in each of the conversational interchanges.
The study shows that for second and foreign language education and education in general, students can profit from acquiring skills that permit them to identify and adjust to different cultural and language-appropriate situations. (L&E 1988-6)

The analysis of the textual structure also considered the relationship to the textual structure of the article itself and the presence of all the moves. Particular attention was paid to the presence or absence of background moves (as identified by Dos Santos 1996). Abstract (3) provides an example of an empirical abstract with the background move 'Situating research':

(3) *How can teacher talk support learning?*

The dialogic nature of discourse which stems from the work of [Bakhtin, M. (1994). Speech genres and other late essays. In P. Morris (Ed.), The dialogic imagination (pp. 81–87). London: Edward Arnold. (Original work published 1986)] provides important understandings in an investigation into the nature of classroom discourse. Using analytical tools informed by Systemic Functional Linguistic theory it is possible to articulate a variety of teacher talk strategies which enable the teacher to apprentice his students into the discipline of History. [Situating research]
By examining in detail *excerpts* from two History lessons that occurred at the beginning of the first year of high school in an independent Australian boys' school, some of the teacher talk *strategies* that lead to students' developing the

> skills and content relevant to a particular subject area *have been identified. These are* repeating, recasting and recontextualising language to develop technical language; cued elicitation; modifying questioning to extend or reformulate student's reasoning and recycling ideas through busy clusters of words. (L&E 2008-7)

After the analysis of the generic structure, the study considered style, looking in particular at the distribution of metadiscourse items in the genre both from a textual and an interpersonal perspective. At the textual level, we classified abstracts according to their general style on the basis of the presence of self-mentions (Hyland 2005). Abstracts were divided into three groups: those that constantly used first person pronouns (personal style), those that constantly avoided them (impersonal) and those that combined the different forms (mixed).

Finally we moved on to a closer study of framework sequences, dealing in particular with the most frequent *verba dicendi*. The analysis – supported by Wordsmith Tools 5.0 – started by exploring the frequency and use of potentially metadiscursive expressions pointing to some of the most interesting metadiscursive nodes of the abstract. Verbs of saying were classified into two main semantic categories, i.e. verbs of claim-making and verbs of topic-setting. The focus then shifted to a close reading of concordances of highlighted elements, looking at collocations, patterns (Hunston/Francis 2000) and semantic sequences (Hunston 2008), thus integrating meaning, form and function with the aim of relating patterns to the structure of the abstract. We also looked more closely at variation in the patterns involving the presence of the researcher, identifying personal patterns (*we argue*), impersonal patterns (*it is argued*) and locational patterns (*the paper argues*) (Dahl 2004).

3. Results

This section is sub-divided into three main parts. In the first section, the changes in the generic structure of the abstract from 1988 to 2008 are discussed. The second section deals with general issues of style concerning metadiscursive phraseology. In the third section, emphasis is laid on differences in the use of framing verbs by comparing patterns of verbs of saying in the two periods selected for the analysis.

3.1. Changes in the generic structure

Considering the quantitative presence of argumentative and empirical research abstracts in the two corpora, we noticed a strong decrease in the first type from 1988 to 2008. Indeed, in 1988, 29 abstracts out of 70 (41%) could be considered as argumentative, while the percentage dramatically decreases in the 2008 corpus, in which we found only 18 argumentative abstracts out of 70 (26%).

It was also noticed that the coincidence between the structure of the abstract and the structure of the abstracted article was generally almost perfect. Only one abstract did not fully account for the content of the article.

Apart from the difference in type outlined above, another significant change occurred in the move structure of the genre over the time span considered for the analysis. When looking at specific moves, 75% of the 1988 abstracts are limited to the four moves identified by Bhatia (1993: 78-79), i.e. *1. Introducing purpose; 2. Describing methodology; 3. Summarising results; 4. Presenting conclusions*. On the other hand, 50% of the 2008 abstracts (9/18 argumentative + 27/52 empirical) present a further background move placed before Move 1 (see Dos Santos 1996), i.e. *Situating research*, in which the author summarizes previous literature about the topic.

To sum up, the 2008 abstracts seem to reflect a higher frequency of research reports with all their sections. Moreover, they usually foreground the basically argumentative nature of academic discourse in a substantial background section (*Situating research*) positioning the research proposed in the academic community. These considerations reflect a focus shift in applied linguistics as a discipline from 1988 to 2008, moving from greater emphasis on discussing theoretical and methodological issues to a more empirical research perspective that gives greater importance to results and discussion.

The presence of the *Situating research* move can lead to a double interpretation: 1) researchers have developed a growing sense of the academic 'community' both as an 'addressee' and as a 'warrant' of the research process; 2) there is a growing competitiveness among researchers that induces them to foreground their claims against previous pieces of research in the field in order to acquire centrality for the topic that will be discussed in the paper.

3.2. Changes in style

Moving on to the presence of the writer's voice in the abstract, we noticed a change in his/her discursive identity, i.e. presence in the text, from 1988 to 2008. The author is more present in the 2008 abstracts through the use of self-reference devices such as nouns and pronouns which almost doubled in the twenty-year time span: the first person pronouns *I/we* increased from 25 occurrences in 1988 to 56 in 2008.

Focusing on metadiscursive phraseology and in particular on *self-mentions* (Hyland 2005), we identified three main styles in abstracts:

1) <u>impersonal style</u>, including four patterns: a) non personal (*The paper sets out to...),* b) impersonal (*It is argued that...),* c) locative (*An attempt is made in the article to infer...*), d) personal 3rd person (*The author suggests...);*
2) <u>personal style</u>, characterized by systematic use of 1st person singular and plural adjectives and pronouns (*I conclude, we argue, my analysis)*;
3) <u>mixed</u>, using both impersonal and personal patterns.

Table 3 presents the results concerning the abstract style in the two corpora, while at the same time providing information on the argumentative (A) or empirical (E) nature of the abstract.

Style	*1988*	*2008*
Impersonal	26 A + 38 E[1] = 54/70 (90%)	12 A + 36 E = 48/70 (69%)
Personal	3 A + 1 E = 4/70 (7%)	5 A + 10 E = 15/70 (21%)
Mixed	2 E (3%)	1 A + 6 E = 7/70 (10%)

Table 3. Abstract style 1988-2008.

As shown in the table, even though the presence of impersonal abstracts is still high (69%), the 2008 researchers seem to signal their presence in the abstract by using personal framework sequences more frequently than the 1988 ones (21% of the 2008 abstracts vs. 7% of the 1988 abstracts). Moreover, whereas in the 1988 corpus the impersonal style seems to dominate the scene (90%), in 2008 we find a more balanced subdivision between impersonal, personal and mixed approaches (69%, 21% and 10% respectively).

3.3. Framing verbs and their patterns

Moving on to the lexical patterns of framing verbs, the analysis of wordlists disclosed the most frequent *verba dicendi* in the two corpora. These were classified into verbs of 'claim-making': *suggest, argue, indicate, explain, find, infer, propose, show, support*; and verbs of 'topic-setting': *examine, discuss, present, analyze, concern, consider, focus, investigate, report, review*.

From a merely quantitative point of view, it is apparent that the trend in using 'claim-making' verbs or 'topic-setting' verbs is diametrically opposed when comparing the 1988 corpus to the 2008 one. This striking difference is well-exemplified in Figure 1.

1 A = argumentative; E = empirical.

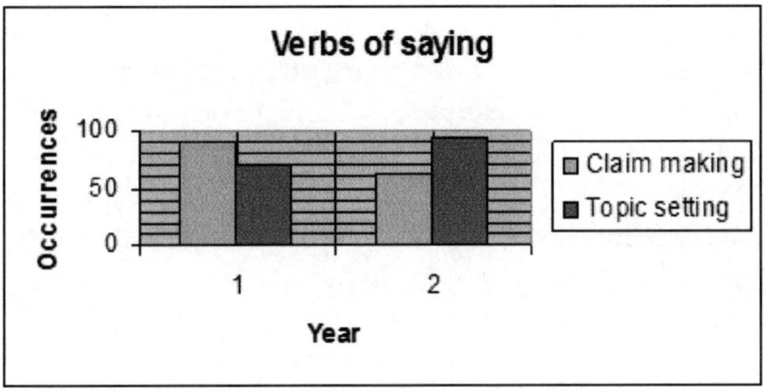

Figure 1. Verbs of saying 1988-2008.

As shown in Figure 1, the occurrences of *claim-making* verbs moved from 90 to 63 from 1988 to 2008. The trend is completely reversed for *topic-setting* verbs which increased from 70 occurrences in the 1988 corpus to 94 occurrences in the 2008 one. This evidence corroborates what has already been said in the previous section (§3.1) about a focus shift in applied linguistics as a discipline from 1988 to 2008. Indeed, these results confirm that the discipline moved from placing emphasis on discussing theoretical and methodological issues to a more empirical research perspective.

After identifying the most frequent verbs of saying and their distribution in the two corpora, we focused on the variation in the presence of the researcher in the abstract and we studied what we call 'personal patterns' (*we argue*), impersonal patterns (*it is argued*) and locational patterns (*the paper argues*) (Dahl 2004). The analytic results are reported in Table 4, showing that the decrease in impersonal patterns is particularly evident for claim-making verbs, while locational patterns double their presence in both classes of verbs:

	Topic-setting verbs				*Claim-making verbs*			
PATTERN	1988	%	2008	%	1988	%	2008	%
Personal	4	5.7	11	11.7	12	13.3	14	22.2
Impersonal	49	70	27	28.7	54	60	10	15.9
Locational	17	24.3	56	59.6	25	27.7	36	61.9

Table 4. Verbs of saying patterns.

When looking at the patterns characterising verbs of saying as a whole, results show a significant increase in *locational patterns* (26% vs. 60%) from 1988 to 2008, a decrease in the use of *impersonal patterns,* i.e. passive forms, (64% vs. 24%) as well as a slight increase in *personal self-mentions* (10% vs. 16%) equally distributed in claim-making and in topic-setting verbs. In the ethos of the community, impersonality seems to have given way to locativity in characterizing the representational work of abstracts.

4. Conclusions

The present chapter aimed at investigating the evolution of the abstract as a genre over a time span of twenty years (1988-2008) considering in particular the field of applied linguistics. Specifically, the study explored the changes in communicative practices and linguistic patterns which may be due to the pressure of technological innovations and the spreading of the genre across wider fields, which has led to writers' greater awareness of its importance.

The significant change in the textual structure of the abstract, in relation to the nature of the article with which it is connected, corresponds to a shift in the focus of the discipline from a predominant need to discuss theoretical issues to a more empirical perspective. To tackle this rhetorical change, the study explored the generic structure of the abstract in the two periods considered and described the main differences in move realization. This shift produced an interesting evolution in the presence of the researcher's voice in the genre, as well as an increasing awareness of the importance of placing one's own research within the debate that characterizes a discourse community.

Data showed that the development of the genre seems to emphasize what Hyland (2000: 63) referred to as promotional features: 1) displaying membership (for example by highlighting expertise in the research process), 2) claiming significance (for example by foregrounding the writer's position in the disciplinary debate, rather than explicit evaluation), 3) signalling 'representational features' by making reference to a displaced text in locational patterns, as well as by referring to the academic context (as reflected in the increased presence of the background move 'Situating research' in the 2008 abstracts).

Impersonality seems to have given way to locativity in characterizing the writer's voice. Locational patterns allow writers a 'middle' way between personalization and impersonality: when talking about their papers rather than about themselves, they can be more explicit in the assessment of their own research, while at the same time retaining a somewhat more objective stance. The combination of locational patterns and references to the community allows writers to strengthen their personal position by reference to collegiality.

DAVIDE S. GIANNONI

A Diachronic Study of Genre Variation in Academic Publishing: *The Quarterly Journal of Economics* (1965-2004)

1. Introduction

Academic journals perform the role of disseminating disciplinary knowledge through a wide range of textual-rhetorical constructs drawn from the genre repertoire of the scientific community they serve. Among these, the most widely-investigated genre is the standard research article, where new findings and/or arguments are presented to peers as an addition to – or improvement on – existing scholarship. A large strand of applied linguistic research has therefore focused on such texts, uncovering the structures, strategies and wordings that enable authors to communicate their findings effectively to fellow-researchers in their field (cf. Bazerman 1988; Swales 1990; Gross *et al.* 2002). In the case of English, which has become the language of choice for international communication in so many disciplines, authors are confronted with the additional challenge of crafting their texts in a way that bridges the gap between local/global (native and non-native) textual norms and expectations.

But journals provide their audience with far more than a mere selection of recent articles. It is sufficient to glance at the table of contents (TOC) of an academic serial to appreciate the diversity of genres represented in its pages, often reflecting editorial efforts to increase the journal's readership and scientific impact. Though not as widely explored as research articles, these 'lesser' genres have attracted the attention of analysts for their pervasiveness and interpersonal significance within domains. Early linguistic descriptions of this kind target review articles (Myers 1991), book reviews (Motta-Roth 1996), article abstracts (Melander *et al.* 1997), editorials (Salager-Meyer 2002) and letters to the editor (Busch-Lauer 2003). Proportionally speaking, however, the literature on academic writing shows a bias towards research articles that deserves to be redressed. Though admittedly due to the emphasis placed on articles as a yardstick for scientific productivity, the role of other genres should not be underestimated. What is more, their use is linked to evolutionary phenomena in academic publishing whose origin can be traced back to 17th century science journals (cf. Atkinson 1999).

The present chapter aims to shed further light on the generic dynamics of English-medium serials by examining a corpus of TOCs and specimen texts published in the world's most influential economics journal over a forty-year period.[1] The methodology developed for this purpose – which comprises both quantitative tools and manual investigation of generic cues – highlights the complexity of studying the diachronic dimension of standard(ising) forms of academic writing. It does not attempt, however, to link such evidence to any grand theory of generic change of the type postulated by Berge (1994). Albeit attractive, the explanatory value of such models (cf. also Vagle 2001) is essentially speculative and cannot be tested without recourse to insider knowledge, ethnographic data and historical details unavailable to the humble corpus linguist.

1.1. Taxonomies and nomenclatures

The consideration of academic texts as exponents of discipline-specific genres is made particularly difficult not by the different conceptualisations of 'genre' produced since the term was first adopted (cf. Hymes 1974), but also for the taxonomical complexities inherent in their classification. The variety of (mostly metaphorical)

---

1 This study is part of a wider project targeting generic evolution in the domains of economics, medicine, law and linguistics. It is supported by COFIN Grant No. 2007JCY9Y9. I am grateful to an anonymous reviewer for his/her comments.

labels adopted for this purpose suggests a considerable degree of uncertainty in the literature, from 'genre system' (Bazerman 1994) to 'genre colony' (Bhatia 1997a) or 'genre network' (Swales/Feak 2000). In a generous attempt to problematise the debate, Swales (2004) suggests using 'genre hierarchy' for functional groupings, 'genre chain' for chronological sequences and 'genre set' for the range of texts employed by an individual. As our concern in this chapter is the presence/absence of certain genres in academic journals rather than their taxonomical structure, the evidence in hand will be viewed in terms of *genre repertoire*. This more neutral term simply identifies "the nature and source of genres that are recognized and accepted by a community of practice as legitimate forms of working and interacting" (Orlikowski/Yates 1994: 23).

The most widely-cited attempt to classify the repertoire observed in written academic discourse is based on a tripartite subdivision into primary genres (intended for public perusal and/or publication), secondary genres (targeting students or novice scholars) and occluded genres (intended for private communication). Table 1 below lists each group, as outlined over the years in the work of Swales (1990, 1996, 2004). While the focus here is clearly limited to entries in the first column, it will be interesting to see which of these are present (or not) in the corpus and whether there are other genres that should be added to the list, whose presence has not yet been duly noted.

Primary	*Secondary*	*Occluded*
Research article	Textbook	External evaluation
Review article	Introductory text	Evaluation letter for tenure
Book review	Post-introductory text	and promotion
Journal abstract	Course description	Book/grant proposal
Editorial	Lecture notes	Review of articles
Monograph	Handout	submitted to journals
Conference abstract		Research grant proposal
Dissertation		Initiating/responsive email
Thesis		Application, invitation,
Chapter in edited book		request, submission and
Case report		editorial-response letters

Table 1. Grouping of written academic genres.

Valuable information as to a journal's genre repertoire may be gleaned from its TOC, where generic labels are often embedded in titles or used to frame certain sections of the TOC. Observed diachronically, the rise or demise of such labels may point to a variety of phenomena, the most notable being:

- genre emergence (through the introduction of new or derived genres);
- genre loss or erosion (when non-functional genres are phased out or encroached upon by alternative options);
- generic hybridity (the merging of features/functions previously associated with distinct genres).

There is an evident link between the three phenomena, as the demise of one genre tends to originate (or be due to) the rise of another, which in turn may be well established or the result of a hybridising adaptive process. The generic ecology of a successful journal, as it competes for visibility in its field, is particularly prone to change; this makes rhetorical 'cross-fertilisation' (Bhatia 2004) the norm rather than an exception in academic as well as professional communication (cf. Giannoni 2006).

2. Material and methods

As the analytical perspective taken here specifically targets diachronic variation in the genre repertoire of academic publishing, the material assembled for this purpose is drawn from a top-ranking journal whose digital archives allow inspection over a minimum of four decades. The field considered (economics) is the first of the four domains chosen for investigation, as part of my contribution to a national research project on 'Tension and Change in English Domain-Specific Genres'. To identify the most influential publication in this field, the latest available edition (2008) of Journal Citation Reports for the Social Sciences in the ISI Web of Knowledge database was queried, using impact factor (IF) as a yardstick. IF measures the total number of citations to a journal over two years divided by the number of articles published in that journal during the same period and is widely considered to be a reliable index of academic achievement. By this standard, the world's top journal in 2008 was *The Quarterly Journal of Economics* (QJE), which the publisher's webpage <http://www.mitpressjournals.org/loi/qjec> describes in the following terms:

> The oldest professional journal of economics in the English language. Edited at Harvard University's Department of Economics, it covers all aspects of the field – from the journal's traditional emphasis on microtheory, to both empirical and theoretical macroeconomics.

While QJE's association with an Ivy League university and its uninterrupted publication since 1886 have certainly contributed to its academic standing, available IF data show that the journal operates in a highly-competitive environment and has not always occupied the top slot in recent years:

Year	*Rank*	*IF*
1999	4	4.3
2000	2	4.0
2001	2	3.8
2002	2	4.0
2003	2	4.8
2004	1	4.4
2005	1	4.8
2006	2	4.7
2007	3	3.7
2008	1	5.0

Table 2. Academic standing of QJE in terms of IF (all available years).

At this point, the TOC of the first issue of each volume was downloaded from the journal's archive and stored for analysis. The forty years covered by the resulting material (1965-2004) reach up to the latest digitally-available volume and, albeit limited, contain interesting evidence of generic variation during a period of unprecedented growth and transformation in the world of academic publishing. To avoid skewing the results, special issues and (where possible) those without a TOC were not included in the corpus. The genres represented in each issue were then classified by checking its TOC and the relevant texts; at times (when no genre label was present) this was performed inductively by exclusion.

To further investigate the genre repertoire identified in the journal, I conducted a comparative keyword-analysis of its content. For this purpose an exemplar of each genre was downloaded for each year considered and converted to .txt format for analysis with WordSmith Tools 5.0 (Scott 2007), after removing such peripheral content as page headers, bibliographic information and copyright notes. This allows the analyst to build a linguistic profile of each category of texts, based on lexical types whose presence is unusually (in)frequent in a given genre as compared to the journal's content in general. This part of the corpus comprises a total of 489,063 tokens.

3. Results

3.1. Genre repertoire (analysis of TOCs)

Variation in the genre repertoire can be better understood if grouped by decade. Table 3 below summarises the evolution observed over a 40-year period, with content classified into six categories. All of the genre labels listed in Column 1 are taken from the TOCs and/or title page of each entry. They are all familiar names, with exception perhaps for Recent Papers, a bibliographic section used in earlier volumes of QJE to draw attention to new publications in the field.

The picture that emerges from these figures is quite clear, insofar as it highlights a drastic simplification of the genre repertoire from six different categories in 1965-1974 to only two in 1995-2004 (n. indicates a figure smaller than one and thus negligible).

	1965-1974	*1975-1984*	*1985-1994*	*1995-2004*	*ALL*
Articles	54%	59%	73%	97%	359
Notes	18%	24%	24%	-	86
Comments	14%	9%	1%	n.	33
Replies	7%	8%	n.	n.	21
Recent pubs.	7%	-	-	-	9
Editorials	-	-	2%	2%	5
Total	*137*	*121*	*140*	*115*	*513*

Table 3. Diachronic evolution in the genre repertoire of QJE (1965-2004).

The most noticeable change is the rise of Articles, in proportional terms, from just over half (54%) of content in 1964-1974 to almost its entirety (97%) in 1995-2004; the other genres were gradually phased out in the 1990s. Editorials are the only exception to this trend, having first appeared in 1985-1994 and subsequently become a standard (if minor) presence in QJE. This means that today the journal only publishes Articles and, very occasionally, a few Editorials.

The differences are particularly striking if one looks at the journal's earliest and latest TOCs, reproduced in the Appendix. The comparison suggests that generic diversity characterised QJE since its foundation and, despite variations, remained a feature of the journal until quite recently. Notes/Comments/Replies, for example, still accounted together for 39% of content in 1965-1974 but disappeared completely in the 1990s.

The overall impression, therefore, is that the less 'central' written genres are losing out against the 'competition' of standard research papers. The almost exclusive reliance on Articles as a channel for academic communication may be linked to two major factors:

- *Productivity*. The pressure on journal editors, publishers and contributors to maximise their scientific impact in terms of citations makes Articles the preferred option, because they are more likely to be cited in the literature and ranked as publications in academia. Conversely, the more interactive genres are viewed as peripheral contributions and thus often ignored when measuring productivity.
- *Instability*. The generic instability (and hybridity) of Notes, Comments and Replies, which tend to be used dialogically in

response to an original article, arguably makes them harder to classify for journal editors and readers. This phenomenon is reflected in the changing terminology observed in the TOCs ('Note' or 'Short Paper'; 'Reply' or 'Response') and by the categorisation of such genres ('Comments' and 'Replies' variously indexed under 'Articles' or 'Notes').

3.2. Genre profile (keyword analysis)

As mentioned in the Introduction, our focus here is on the minor genres of academic publishing. For this reason, their 'distinctness' was investigated using comparative keyword analysis and manual investigation of the neighbouring co-text. Keywords are a valuable source of information in genre profiling, as they reveal "not only a great deal about the subject matter, the 'aboutness' of a particular text genre, but it can also help specify the salient features which are functionally related to that genre" (Goźdź-Roszkowski 2010: 72).

The entire collection of QJE texts (98 texts comprising 498,063 tokens) was uploaded to WordSmith Tools as a reference corpus in order to reveal any unusually (in)frequent types[2] in each of the four minor genres identified earlier: Notes (26 texts), Comments (17), Replies (13) and Editorials (5). The results were then approached bearing in mind Baker's (2004: 357) warning that "as with all statistical methods, how the researcher chooses to interpret the data is ultimately the most important aspect of corpus-based research". For the sake of consistency, the analysis was limited to the top six positive and bottom two negative keywords in each genre.

3.2.1. Notes

The purpose of Notes is to present new ideas to peers for evaluation. These are often of a tentative nature and seek to trigger a debate rather than add to the knowledge base. The keywords below are associated

2 Keywords were calculated by log likelihood (min. frequency 3, max. n. 500, max p. value 0.000001).

with various generic features amenable to lexical analysis. Column 2 shows each item's frequency as a proportion of all tokens in the Notes corpus, whereas Column 3 gives its percentage calculated on the global reference corpus.

Keyword	*%*	*% Ref.*	*Keyness*
be	0.82	0.63	35.64
let	0.09	0.04	33.65
where	0.23	0.14	31.95
may	0.29	0.19	30.68
therefore	0.10	0.05	25.94
my	n.	0.04	-37.70

Table 4. Genre-specific keywords in Notes.

Infinitive *be* (648 occurrences) was found above all in passive constructions, and less often with predicative *found* (18), *shown* (18), *used* (11) and *written* (18). Its top L-collocates are all epistemic modals reflecting the genre's tentative, speculative nature: *can/cannot* (141), *may* (104), *will* (91), *would* (49), *should* (40):

(1) The market value of the firm must not fall below a certain proportion v(≤1) of the value of its assets; otherwise there *would be* the risk of a takeover bid. N1979

The keyword *let* (70 occurrences) occurred mostly in the expression *Let* X + VERB and suggests that, compared to other genres, Notes favour the use of mathematical arguments. In 22 instances it was part of the metadiscoursal form *let us:*

(2) *Let* p and q *denote*, respectively, the world and domestic relative prices of imports in terms of exports, so that q = p (1 + t) if t is the tariff rate. N1984

(3) *Let us* suppose that the subset of efficient pairs belonging to {y, P_1} is strictly convex. N1973

Also *where* (180 occurrences) tends to be embedded in a crystallised expression – i.e. *Where* X *is/are* (85) – which confirms my previous observation on the use of mathematical arguments in Notes.

(4) It is calculated as y = X exp((-0.004167)(Z)), *where y is* experience in months, X is length of time of longest job in months, and Z is months since longest job. N1983

The modal *may* (227 occurrences) foregrounds epistemic tentativeness; at times it is found in the (reader-inclusive) metadiscoursal form *we may*.

(5) An unweighted index of linkage *may* fail to depict the pressure-creation ability of the various industries, for its calculation procedure assumes that all industries are of equal importance in a system of industries. N1976

The last positive keyword, *therefore* (78 occurrences) is an argumentative connector linked to mathematical or general reasoning; it occurs sentence initially in less than half (30) of instances.

(6) This paper *therefore* provides further support for the need to integrate the two approaches to the study of profits as described by Caves and Yamey. N1977

The only negative keyword was *my* (5 occurrences), whose infrequency indicates avoidance of first-person attribution. As the proportion of single-authored texts (74%) is identical in Notes and in the global corpus, the former are somehow unwilling to assert ownership of claims, perhaps again because of the genre's tentative (and potentially controversial) content.

3.2.2. Comments

The purpose of Comments is to provide evaluation of peers' work, contrasting it with other research (often including the authors' own data). This may lead to direct confrontation and escalate into open conflict. The table below lists the genre's distinctive keywords.

Keyword	*%*	*% Ref.*	*Keyness*
would	0.42	0.20	61.08
professor	0.08	n.	60.37
was	0.31	0.14	53.98
article	0.07	n.	48.51
comment	0.08	0.01	47.91
our	0.03	0.13	-37.42
we	0.17	0.42	-69.92

Table 5. Genre-specific keywords in Comments.

The speculative conditional *would* (157 occurrences) is used to engage the interlocutor in a discussion, its most prominent R1-collocates being *be* (143), *have* (17), *increase* (6) and L1-collocates *one* (7) and *it* (6).

(7) To contradict this proposition, which is central to the Marxist theory, one *would have* to show that the hierarchical organization of work [...] is not integral to capitalist profit-making. C1978

Another interesting keyword is *professor* (29 occurrences). This honorific is used almost exclusively as a title for peers targeted by the author's (often critical) comments; at times the combination produces a clearly ironic effect. Only three realisations lacked the peer's proper name.

(8) *Professor Robinson's* account revives some of my own memories. [...] For, as *Professor Robinson* says, the purely logical parts of economic arguments should not remain subject to irresolution or dispute. C1975

The prominence of *was* (114 occurrences) foregrounds the genre's focus on past claims/actions and their relative value. It includes 13 negative evaluations involving *erroneous, inadequate, inconsistent, merely, never, not, no, unable, too*.

(9) Ohlin [1933, p. 563] tells us that he *was* led to re-examine international economics from the mutual equilibrium (or market price) perspective because conventional theory *was unable* to explain currency price variations from purchasing parity during the immediate post-war period. C1981

Linked to evaluation is also the item *article* (24 occurrences), whose main L-collocate is *in* (11). It is often qualified possessively and signals almost exclusively the target of the author's comment.

(10) *In their article,* 'The Effects of Taxation on the Selling of Corporate Stock and the Realization of Capital Gains', S. Feldstein and Shlomo Yitzhaki [1980] provide a quantitative analysis of a vitally important subject. C1984

Of special interest is the genre-label *comment* (28 occurrences), a metadiscursive deictic item observed nearly always in the text's title or opening paragraph; its most common L1-collocate is *this* (9).

(11) *In this comment* issue is taken with the explanation offered for the observed positive relationship between the proxy for the extent of nationalism [2] and per capita income. C1967

Also two negative keywords, both first-person plural markers (*our*, *we*) were found in the corpus. Their rarity, however, is merely due to the fact that only 1/17 of Comments have multiple authors, compared to 31% of texts in the global corpus. In the case of *we* (61 occurrences) half of all instances encode reader-inclusive (occasionally polemical) claims.

(12) If *we* examine the economies in which footloose industry has flourished, we find generally small, export-oriented countries in which the structure of production conforms rather well to existing factor endowments. C1976

3.2.3. Replies

As the name implies, Replies are published in response to controversial contributions published in the same journal (generally, but not always, in an earlier issue). The fact that they address a specific text and author(s) is reflected in the intensely personal nature of their claims – an aspect confirmed by the presence of three possessive adjectives among the genre's top six keywords. More specifically, *my* is the item with the highest level of keyness (75.72) identified so far in the corpus and the figure is significant given that the proportion of single-authored texts (75%) is almost identical to that of the global corpus (74%).

Keyword	*%*	*% Ref.*	*Keyness*
my	0.23	0.04	75.72
his	0.27	0.08	58.17
reply	0.08	n.	55.97
our	0.35	0.13	51.20
theoretical	0.10	0.01	43.28
believe	0.09	0.01	42.00

Table 6. Genre-specific keywords in Replies.

In line with such orientation, *my* (50 occurrences) contrasts the author's work with that of his/her target – its most common R-collocates being *study* (7) and *results* (4):

(13) Professor Samuelson is kind enough to make me a number of compliments in his comment. I would be more gratified if he would answer *my* point. R1975

In at least one instance, the (female) author's self-promotional drive produces a realisation that betrays an uncommon lack of academic modesty:

(14) Both *my intuition and ten years of work in criminal justice research* lead me to believe that both sanctions and opportunities affect criminality. R1983

The second keyword (*his*, 60 occurrences) identifies the target, who is typically male, as the author/originator of rival knowledge claims. Its top L-collocate *in* (23) is associated with such items as *analysis* (8), *comment/s* (4), etc.

(15) Myers' specification in his second analysis contains many fewer 'taste' variables than either *his own first analysis* or my analysis. R1983

As in the case of Comments, the genre label (*reply,* 17 occurrences) is a keyword also here. It tends to occur in titles, with only three exceptions as in example (16) below:

(16) In this *Reply* I shall do two things. R1983

First-person attribution is particularly prominent also in the case of multi-authored texts, with 77 occurrences of *our*. This R-collocates especially with *estimate/s* (14), *paper* (10) and *interpretation* (4).

(17) In *our paper* we estimate that the change in 1973 to a system that capped the effective tax rate at 0.25 would induce additional net realizations of $17 billion annually. R1984

The next keyword (*theoretical*, 22 occurrences) has *framework* as its main (22) R1 collocate. This arguably suggests a propensity of Replies to engage with theoretical issues more often than other journal genres:

(18) Even if the above criticisms on the *theoretical framework* can be met, there appear to be additional difficulties with regard to the parameters used in the application of this framework. R1980

Finally, the prominence of the relatively strong cognitive verb *believe* (20 occurrences) mostly associated with the L-collocate *I/we* (14) confirms the highly assertive nature of Replies:

(19) However, *we believe* that the procedure developed by Hausman and Wise [1981] to correct the bias is not appropriate for this particular application. R1984

3.2.4. Editorials

Editorials provide a venue for institutional communication between the journal's editor(s) and its readership. Their presence is linked, at least in part, to the increasing tendency to foreground the personal element in scholarship (cf. Morris 1988; Giannoni 2008a, 2008b). The genre first surfaced in QJE in the mid 1980s as an introductory 'Manifesto' whose opening lines read as follows:

(20) With this issue a new Editorial Board of the QJE makes its debut. We have assembled an impressive group of Associate Editors whose names appear on the masthead. E1985

Albeit limited to five texts, the Editorials section of the corpus had several interesting keywords that distinguish it from other types of journal content.

Keyword	%	*% Ref.*	*Keyness*
editorial	0.88	n.	84.11
editors	0.78	n.	71.42
accepted	0.59	n.	42.82
board	0.59	0.01	33.11
publish	0.29	n.	26.13
will	1.27	0.21	24.81

Table 7. Genre-specific keywords in Editorials.

The word *editorial* (9 occurrences) has the highest keyness value observed in this analysis. Contrary to expectations, however, it is always used adjectively to qualify the QJE's *board* (3), *data* (2), *decision* (3) or *speed* (1). Similarly, the keyword *editors* (8) identifies past or present members of QJE's editorial board. This is the only group of texts where multiple authorship is the norm rather than the exception. In such realisations the writers tend to distance themselves rhetorically from the authors through third-person reference:

(21) This issue of the Quarterly Journal of Economics marks the retirement of Olivier Blanchard and Andrei Shleifer from their service as editors. The new *editorial board* (Alberto Alesina, Edward Glaeser, and Lawrence Katz, who is continuing as editor) would like to pay tribute to the work of these two exceptional *editors*. E1999

Another keyword reflecting the function of Editorials is *accepted* (6 occurrences), which always collocates with *manuscripts/papers* submitted to the journal:

(22) The following *manuscripts*, in addition to those listed in previous issues, have been *accepted* for publication in forthcoming issues of The Quarterly Journal of Economics. E1986

The institutional term *board* (6 instances) co-occurs, predictably, with *editorial* or *of editors*, while *publish* (3) identifies QJE's editorial activities.

(23) We intend to *publish* larger volumes in the future whenever necessary to keep the time between acceptance and publication at a maximum of six months. C1985

Finally, the presence of *will* (13 occurrences) reflects the genre's orientation towards future events (e.g. a *prize/award*) or editorial decisions. In one of the three instances introduced by *we* it also signals volition:

(24) We salute Andrei and Olivier for their wisdom, their effort, their intellectual honesty, and their bravery. *We will* do our best to ensure that the Journal continues to embody their spirit of intellectual discovery and scholarly excellence. E1999

The diversity of functions performed by Editorials is highlighted by the range of genre labels observed in the corpus: namely, 'Announcement' (2 texts), 'Accepted Manuscripts and Editorial Data' (2) and 'Manifesto' (1).

3.3. Functional interpretation

The keywords identified so far as indicative of the generic profile of the four genres analysed in this study can be assigned to three large categories, according to their function.

- *aboutness* for those linked mainly to a genre's content (ideational function);
- *style* for variants that reflect the argumentative style favoured in a genre (textual function);
- *stance/modality* for keywords related primarily to the negotiation of knowledge claims within the discipline (interpersonal function).

While the first two categories were first suggested by Scott (1999c) and are well-documented in various corpus studies (cf. Ljung 2002; Baker 2004; Freddi 2005) of genre-related variation, the third is introduced here to account for all those keywords whose function is primarily interpersonal, insofar as they allow writers "to create a recognizable social world through rhetorical choices which allow them to conduct interpersonal negotiations and balance claims for the significance, originality and plausibility of their work against the convictions and expectations of their readers" (Hyland 2005: 176).

Table 8 below summarises the results using this framework (with negative keywords in square brackets). The overall impression is that aboutness is the most distinctive function in Editorials, style in Notes and stance/modality in Comments and Replies.

	Aboutness	*Style*	*Stance/Modality*
Notes	-	be, let, where, therefore	may, [my]
Comments	article, comment	was	would, professor, [our], [we]
Replies	reply, theoretical	-	my, his, our, believe
Editorials	editorial, editors, accepted, board, publish	-	will

Table 8. Keywords classed by genre and function.

4. Discussion and conclusions

In this exploratory study of diachronic variation in academic publishing, texts covering a 40-year period were extracted from the archive of a leading economics journal (QJE) in order to compare their range, based on the nomenclature employed and on a keyword analysis of generic traits. The special significance of genre labels as a source of linguistic information stems from the fact that scholars "give genre names to classes of communicative events that they recognize as providing recurring rhetorical action", so that "particular attention therefore needs to be given to the genre nomenclatures created by those who are most familiar with and most professionally involved in those genres" (Swales 1990: 55). The results point to considerable quantitative/qualitative differences in the genre repertoire sanctioned by editorial gatekeepers in response to the readerships' evolving communicative needs, since striking the right balance between change and continuity is a major factor contributing to a journal's scientific standing in its discipline.

Generic diversity has been dramatically eroded over the years, with Articles becoming almost the only genre represented in QJE since the mid-1990s. As a result, the more dialogic/evaluative genres such as Notes, Comments and Replies have gradually disappeared from the journal. This trend may be partly accounted for by the pressure on editors to maximise the impact factor (which relies almost entirely on articles) and to the rhetorical instability of the other genres, whose hybridity is confirmed by a considerable degree of taxonomical fuzziness in the TOCs considered. The only exception is the sporadic presence of Editorials (first observed in 1985).

Against such a background, the keyword data extracted from these non-central genres provides interesting evidence of their communicative purpose and linguistic profile. A representative selection of texts covering the whole genre repertoire of QJE was used as a reference corpus. Most of the words identified point to the ideational (aboutness) or interpersonal function (stance/modality), while fewer reflected their textual function (style). Editorials were the most marked for aboutness, Comments and Replies for stance/modality, and Notes for style. The intersection of these three functional planes accounts for the rhetorical distinctness of each genre, regardless of its length or specific topics.

The overall impression, therefore, is of a growing distance between research articles (which produce more citations) and minor contributions, which enrich the debate but are less quoted. However it remains unclear whether journal editors actively promote this shift or are simply bowing to external pressure. Whatever the case, it can be argued that genres "package information in ways that conform to a discipline's norms, values, and ideology" (Berkenkotter/Huckin 1993: 476) not only internally but when they come together as a group within the journal's genre repertoire. Editorial decisions are thus a source of crucial influence for diachronic evolution in academic settings, as they promote certain genres and phase out others. This consideration brings us back to the heart of the problem and to Bhatia's (1997b) argument that in a disciplinary community generic change is largely a result of the 'power and politics' of its institutional members.

Further research is of course needed before the insights presented here can be reliably generalised. This principally means extending the analysis to other top-rated journals/disciplines whose archives allow investigation of the same 40-year period. What I hope to have shown is that genre variation can be fruitfully gauged through a combination of quantitative (keyword extraction) and qualitative tools (manual investigation of data) targeting non-structural features of the genre repertoire employed in modern-day academic publishing.

Appendix

TOCs for February 1900 and February 2004

STEFANIA M. MACI

Poster Makers Should Think as Much about Show Business as Science. The Case of Medical Posters in a Diachronic Perspective*

1. Introduction

According to Maugh (1974) poster sessions were pioneered in Europe and made their first appearance in the US at the Biochemistry/ Biophysics 1974 Meeting. Since then posters have rapidly become a major format for scientific communication at conferences and other scientific meetings (Matthews/Matthews 1996: 97). The purpose of a scientific poster is to share recent research findings with peers, thanks to the development of constructive discussion about a study's research design and results; in doing so, poster presenters can improve their research (Demarteau *et al.* 2007: 91). Indeed, ever since the end of the 1980s, there has been the recognition that medical knowledge can be absorbed more from posters than from oral presentations. For this reason, posters are required to be prepared by focussing more on concrete messages, evidence, and (helpful) titles than on traditional scientific communication, usually based on the contextualization of the study within the relevant literature, the description of scientific protocols to be applied to the research, the report of the findings resulting from the application of the protocol, and the authors' argumentative discussion aimed at persuading the readership about the validity of the whole procedure. Perhaps more than any other form of scientific communication, a successful poster combines visual, oral,

* The research on which this study is based is part of the MIUR-funded research project of national interest (PRIN) *Tension and Change in English Domain-Specific Genres* (COFIN grant no. 2007JCY9Y9).

and written elements. Indeed, a poster presentation is mainly a visual display of research key points which are highlighted by the author(s) in an interaction with the audience. In order to do so, poster presenters have to attract the readership's attention by means of colourful titles, clear tables and graphs, and as few words as possible. In other words, "poster makers should think as much about show business as science" (Whimster 1989: 274). Because poster sessions have evolved so rapidly, their format is still more flexible than better-established forms of scientific communication.

The impact of a changing and more business-oriented health care system on medical education seems to be responsible for certain profound changes in academic medicine. Doubtlessly, competition, financial constraints and medical worth may raise conflicts within the medical discourse community, which are likely to result in generic tensions. Posters seem to be affected by these tensions, which are apparently resolved through the marketization process characterizing medical discourse in general. In the context of a very competitive market, the ability to attract audience is an indicator of product success. In the medical academic field, which is highly competitive and nowadays greatly affected by corporate mentality (Fairclough 1993), peers are attracted by means of strategies typical of promotional discourse.

Although there are a lot of scientific guidelines as to poster preparation and presentation,[1] the study of research posters from a linguistic angle has been carried out only in recent times and only in descriptive terms (Swales/Feak 2000; Burgess/Fagan 2004; MacIntosh-Murray 2007), despite the fact that medical discourse has been analysed from various perspectives. To the best of my knowledge little attention has been paid to the investigation of the relationship between the socio-financial context and medical discourse and to elements which are clear indicators of marketization processes within such discourse. It is therefore the aim of this chapter to investigate such relationship with

1 Scientific guidelines about research posters deal with recommendations as to fonts, colour, size and poster macrostructure. See, in particular Block (1996), Stanojeviæ-Bakiæ (2002), Keegan/Bannister (2003), Di Blasio (2004), Smith *et al.* (2004), AIFA (2005), Mitrany (2005), Erren/Bourne (2007); Lorenzoni *et al.* (2007), Miller (2007), Willet *et al.* (2008), De Castro (2009) and Ellerbee (2009).

particular regard to the discourse of medical posters where marketization processes seem to be more evident, given the consistent presence of visual promotional traits within such a genre (Maci 2011). The research questions at the basis of such investigation will be:

a) What are, if any, the linguistic features within medical discourse that affect poster presentations?
b) What direction have these features taken across time?

A diachronic investigation will be carried out in order to detect the processes undergoing the development of corporate mentality in medical discourse and the extent to which traits pertaining to promotional language have contaminated contemporary discourse in the medical academic community. Although the most evident *marketized* element of a poster is its visual component, for the purpose of this study attention will be paid only to those linguistic features that pertain to medical argumentation and persuasion. Since persuasion exploits the use of evaluative language, I will analyze the keyword lists and the relative collocation patterns generated from three corpora[2] corresponding to the three decades (i.e., 1980-1989, 1990-1999, 2000-2009) in which the posters forming each corpus were published in major medical journals. The discussion of the relative results will be carried out in the following paragraphs.

2. Corpus collection

The main problem I had in order to carry out this research was related to the selection of posters. The Internet offers a myriad of websites regarding scientific poster publication. Indeed, a search on Google using the phrase "medical poster publication" has yielded 7,270,000 hits with topics ranging from tips and techniques, to discussions of how to use posters as a tool for professional development in the

2 See the Appendix for a list of posters.

workplace and to posters themselves. The large number of Web-sites about posters indicates the widespread role posters have in the dissemination of academic medical knowledge, but such overload of information implies that reliability must be the main parameter in poster selection and corpus collection. I therefore checked on the *Journal of Citation Reports*[3] all medical journals, ranked according to their Impact Factor (henceforth IF), and browsed them in order to find publications of poster sessions presented at medical conferences, starting my quest from 1974, the year in which news about a poster session was first published. The journal which first started publication of poster sessions did so during the 1980s and was specialized in the epidemiological field (though it stopped poster publication in 1994); I therefore decided to concentrate on epidemiology and look for posters reporting research involving epidemiological matters. I thus collected 2,638 posters, that is, all available posters published in the major journals of epidemiology or dealing with epidemiological issues, forming a corpus of 1,410,058 words. All posters were then grouped into three sub-corpora according to the decade of publication, i.e. 1980-1989, 1990-1999, and 2000-2009, as indicated in Table 1 below.

1980-1989		*1990-1999*		*2000-2009*	
Posters	Words	Posters	Words	Posters	Words
350	97,858	398	315,233	1,890	996,967

Table 1. Breakdown of corpus composition.

The different size of the three corpora may pose some problems as to the corpus-based investigation of lexical keyness. Creating three corpora of relatively equal size would have doubtlessly been less problematic but it would have meant to skew any diachronic interpretation of my corpus-based approach. Therefore, results have been standardized in Type/Token Ratio (henceforth TTR).

3 See <http://thomsonreuters.com/products_services/science/science_products/a-z/journal_citation_reports>.

3. Preliminary results and discussion

The ability to communicate efficiently and effectively is at the basis of the success or failure of many businesses, and medical institutions are no exception. Miscommunications in the spread of medical knowledge can lead to erroneous diagnoses, detrimental surgical errors, and many other anomalies. An understanding of medical discourse helps to emphasize the verbal and writing skills needed to succeed in the scientific academic community. There are three main communication forms between medical personnel: written, oral, and technological. Yet posters, as stated by Swales (2004: 21), are a hybridization of these three communication forms. In this sense, the poster may be represented as a multimodal communicative event (cf. Kress/van Leeuwen 2004), with text, graphics, colour, and (interactive) speech used to convey meaning. Its content is displayed as a visual unit on a single view plane, which distinguishes it from the research paper (MacIntosh-Murray 2007: 352). In other words, posters are a means to *show* medical research. Nonetheless, the most common way to communicate medical knowledge is through writing; therefore the writing category and writing skills are invaluable to the medical community.

Written medical discourse is presented to readers in a highly codified structure that transcends national cultures (Dahl 2004: 1822). Indeed, such a structure has a set of constraints that have been described in the *Uniform requirements for the manuscripts submitted to biomedical journals* realized in 1978 and known as the *Vancouver Style*, published 1991 in the *British Journal of Medicine* (*BJM*) by the International Committee of Medical Journal Editors (ICMJE), whose updated version (Dec 2009) is downloadable from <http://www.icmje.org/urm_main.html>. Nowadays, scientific posters follow the same organizational conventions as other scientific writings, with an Introduction (and Objectives), Methods, Results, and Discussion/Conclusions. It is the latter of these sections that, traditionally, is regarded as the most relevant in medical discourse by the scientific community as it is here where the paper "is sold" (Docherty/Smith 1999: 1221) and where communication seems to acquire traits belonging to promotional language. It is here that the author's research is advertised and it is here that the reader is persuaded. The *Discussion* section testifies to the strategic use of language to convince the reader of the scientific validity and generalisability resulting from the statistical analysis of data (Horton 1995: 310). In posters, however, space constraints do not facilitate the exploitation of rhetorical devices to emphasize one viewpoint rather than another so as to highlight the quality and limitations of the methods, the clinical importance of the research, speculations about the meaning of the results, directions for future research, as well as claims about the success of the original purpose of the study. Posters, therefore, mould a well-defined structure with persuasion being realized by means of different strategies rather than traditional rhetorical tools.

My first assumption was then that a lack of discursive persuasiveness must be compensated by lexical items carrying evaluative weight. Such lexical items might have a special position within the marketization process of medical discourse. I thus created for each corpus a keyword list generated by comparing the frequency list of the three sub-corpora to a corpus of medical texts of 1,264,074 words comprising abstracts, research articles, clinical studies, reviews, editorials, commentaries and research letters published between 2000 and 2010 by *Jama, The Journal of the American Medical Association (Jama)* (IF 28.89), *The Lancet* (IF 29.44), *The New England Journal of Medicine* (IF 47.05), *Circulation* (IF 14.81), and the *European Heart Journal* (IF 8.9) in order to see if and to what extent the priority assigned to a certain item reflects corporate mentality and how such priorities have changed over a thirty-year time span.

My analysis has required the cross-checking of each word with its collocation pattern in order to decide whether the word was intended with a medical connotation or not. All domain-specific words have been deleted. The rationale of this procedure can be accounted for by the fact that domain-specific terms are technical terms with no evaluative connotation. The resulting keyword lists therefore comprise non-technical words carrying a non-domain-specific meaning. For the purposes of this study, I decided to analyse only the first 30 keywords out of the 500 keywords generated for each corpus with Wordsmith Tools 4.0 (Scott 2004) as shown in Table 2 below:[4]

1980-1989 Sub-corpus		*1990-1999 Sub-corpus*		*2000-2009 Sub-corpus*	
N	*KEYWORD*	*N*	*KEYWORD*	*N*	*KEYWORD*
1	women	3	were	4	public
2	age	6	cases	7	social
3	cases	7	age	9	conclusions
4	risk	8	and	11	methods
5	years	12	aged	12	aim
6	aged	13	years	14	background
7	for	14	CI [Confidence Interval]	15	details
9	smoking	17	women	16	among
10	were	18	of	17	suppl.
14	CI [Confidence Interval]	22	was	18	the
15	males	24	odds	19	and
16	body [premodifier]	26	controls	21	Introduction
17	among	27	smoking	23	socio
19	controls	31	alcohol	24	results
20	white	32	audit	30	was
21	and	33	survey	31	self
22	rates	31	factors	32	were
23	factors	37	questionnaire	38	population
25	smokers	38	authority	39	objective
27	persons	41	social	40	alcohol
29	odds	43	H [hour]	41	contact
31	Dept.	46	risk	44	prevalence
32	alcohol	47	records	45	students
34	authors	50	occupational	47	smoking
36	relative	51	aim	48	apache
37	index	53	services	50	behaviour
40	population	54	collected	51	years
41	race	55	smokers	52	conclusion
42	mothers	56	rates	53	university
43	females	57	residents	55	questionnaire

Table 2. Keyword lists of poster corpora.

4 The numbers in the columns whose heading is N refer to the position of the term as keyword (the missing numbers and relative words are medical-domain terms).

As we can see from the three lists, there is apparently variation in keyness from the various decades. Indeed, the only lexical items common to the three sub-corpora are *and, alcohol*, *smoking*, *were* and *year*. In all sub-corpora, the conjunction *and* semantically implies union and comparison; the words *alcohol* and *smoking* seem to suggest a conscientious recognition of health problems either deriving from *smoking* or strictly *alcohol*-related, as we can see in the excerpts below:

(1) To provide additional data on the smoking-breast cancer association [...]. (P3_1987)

(2) [...] (crowded housing in childhood, low body height, overweight, risk use of alcohol, smoking, blood pressure) was collected among 49 321 men, born in 1949 [...]. (P203_2008)

The word *were* is used in all sub-corpora as an auxiliary to form the past passive in 63% of the cases (out of 982 occurrences); and the term *year* refers to the time-span necessary for the follow-up analysis. In the following paragraphs, I will illustrate the keyness values characterizing the three sub-corpora.

3.1. The 1980 and 1990 sub-corpora

As we can see from the table above, the corpora of the 1980s and 1990s share the majority of lexical items, and precisely: *for*, *women*, *age*, *aged*, *risk*, *cases*, *aged*, *CI* (confidence interval), *factors*, *controls*, *rates*, and *odds*. These are mainly at the top of the keyword list in both decades, which may suggest respect for both lexical conventions and tradition within medical discourse. These shared terms have different keyness values in the 1980s and 1990s corpora, as we will see below.

In the 1980-1989 corpus attention seems to be focussed on *risk*, a word ranked as having high keyness, and on statistical analysis based on categorized *factors* such as, for instance, *age* and *sex* (*male* vs *female*) and supported by *CI*[5] and *odds*, which may indicate that the

5 In statistics, *CI* is the confidence interval within which the estimate parameter taken into consideration and applied on a population sample seems reliable.

experiment to determine and/or eliminate the risk has been statistically tested:

(3) […] a Poisson rate analysis which estimates the relative risk at different exposure levels. (P5_1980)

(4) [...] compared with a black livebirth (95% CI = 0.3-1.3), and for smokers, the odds ratio was 3.6 for a white livebirth compared with a black livebirth (95% CI). (P27_1987)

In the 1990-1999 corpus, attention seems even more focussed on:

(i) statistical computations (for instance, *CI*, and *odds* have a higher keyness value than they had in the previous decade), which is more clearly indicated in the Concordance list (1) below:

Concord

File Edit View Compute Settings Windows Help

N	Concordance	Set	Tag	Word #	t. #	os	#	os	#	os	t. #	os
1	medium SBP have lower refusal rates: odds ratios (OR) 0.28 (95% CI 0.24,			22 659	958	6%	0	2%			0	2%
2	proportion of deaths at home (odds ratio 0.65, 95% CI 0.58, 0.70)			19 135	802	5%	0	8%			0	8%
3	smoked cigarettes: test for trend in odds ratio p=0.009 and p=0.02			9 634	390	5%	0	9%			0	9%
4	produced statistically significant odds ratios for many of the relationships			14 959	633	0%	0	3%			0	3%
5	risk of stroke-age and sex adjusted odds ratios 0-33 (95% CI 0-2, 0-6) and			8 581	376	6%	0	2%			0	2%
6	middle quartiles-age and sex adjusted odds ratios 2-1 (95% CI 1-2, 3-9) and 2-1			8 507	373	8%	0	1%			0	1%
7	population. Main outcome measures - Odds ratios (relative risk) for stroke by			8 382	369	9%	0	1%			0	1%
8	with increased risk for lung cancer (odds ratio (OR) = 4.4, 95% confidence			26 492	705	7%	0	8%			0	8%
9	an increasing linear trend in the relative odds for all lung cancer with decreasing			25 359	635	8%	0	4%			0	4%
10	women with FH (fourth quartile odds ratio (OR) = 0.02, 95% confidence			24 791	600	8%	0	2%			0	2%
11	higher risk for basal cell carcinoma (odds ratio (OR) = 4.0, 95% confidence			24 137	547	3%	0	0%			0	0%
12	misclassification shifts the observed odds ratio toward its null value, and may			23 969	537	7%	0	9%			0	9%
13	associated with lighter skin (odds ratio (OR) = 3.2, 95% confidence			23 679	521	8%	0	8%			0	8%
14	repair levels that significantly increased odds ratios were independently			23 671	521	5%	0	8%			0	8%
15	for aspirin (relative to mild users, the odds ratios (ORs) were 0.94 for			22 155	417	5%	0	2%			0	2%
16	confidence interval (CI) 0.6-7.2). The odds ratio for cancer in the family was			21 912	398	1%	0	1%			0	1%
17	last residence before diagnosis. The odds ratio for having lived within 2.6			21 892	397	4%	0	1%			0	1%
18	were more likely to be premature (odds ratio (OR) = 2.6, 95% confidence			20 840	334	3%	0	7%			0	7%
19	smoke/fire (1.9). Young maternal age (odds ratio (OR) = 9.0, 95% confidence			20 343	304	0%	0	6%			0	6%
20	regression was used to estimate odds ratios and 95% confidence intervals			20 295	302	1%	0	5%			0	5%
21	more bother for at least one symptom (odds ratio (OR) = 1.87, 95% confidence			18 664	206	2%	0	9%			0	9%

Concordance list 1. Concordance list of *odds* in the 1990-1999 corpus.

Such focus on statistical analysis is also evident from the terms *survey* and *questionnaire*, whose resulting data can be compared to establish the association between different types of findings, as we can see in excerpts (5) and (6):

(5) The aim of this particular study was to investigate three factors of questionnaire design in relation to the rate of return and the total number of […]. (P3_1990)

(6) [...] of smoking and drinking. Results - A total of 8 1 6 adolescents completed the questionnaire. (P4_1994)

(ii) Attention is also paid to clinical *audit*, i.e. the systematic analysis of the quality of healthcare system:

(7) [...] Such changes would not have been found in a retrospective audit. The main disadvantage of the prospective audit was the time taken [...]. (P90_1993)

In other words, the stress on statistical computation and systematic analysis in the 1990s sub-corpus, when compared with the corpus of the previous decade, seems to favour the tendency to systematize scientific research whose data, consequently, appear as expressed through well-defined and precise medical discourse.

In the 1990-1999 subcorpus, one of the keywords is the term *aim*. This presence is meaningful as it seems to indicate a new trend in medical discourse. The frequency list generated in the (macro)corpus comprising the three decades, indicates that it is in the 1990s that the lexical items *Introduction*, *Methods, Results* and *Conclusion* begin to be used as headings and acquire the role of section marker within the poster presentation.

As we can see in Table 3 below, in the 1990s for the first time ever *Results* appears with 93 hits (TTR 0.29) as a heading with this section-marker function, alongside with *results* as a synonym of *findings*, which however occurs more frequently (263 hits, TTR 0.83). Furthermore, in some cases *Results* stands for *Discussion* since the latter occurs only 16 times (TTR 0.05) in the 1990-1999 corpus; it is followed by the lexical item *Conclusion** (87 occurrences, TTR 0.27), in which the final comments on the research are stated. A comparison of the occurrence of the common Introduction-Methods-Results-Discussion (henceforth IMRD) amongst the three corpora can be seen in Table 3 below:

		1980s	*1990s*	*2000s*
Introduction		0	17 (TTR 0.05)	617 (TTR 0.61)
Objectives	*Objectives*	4 (TTR 0.04)	37 (TTR 0.11)	137 (TTR 0.13)
	Aims	0	47 (TTR 0.14)	2 (TTR 0.002)
Methods		0	212 (TTR 0.67)	1468 (TTR 1.47)
Results		0	93 (TTR 0.29)	1655 (TTR 1.66)
Discussion	*Discussion*	1 (TTR 0.01)	16 (TTR 0.05)	57 (TTR 0.05)
	*Conclusion**	3 (TTR 0.03)	87 (TTR 0.27)	1621 (TTR 1.62)

Table 3. Breakdown of the IMRD pattern among the three corpora.

3.2. The 2000s sub-corpus

The generic constraints required by the ICMJE since 1993 have clearly influenced the linguistic pattern of posters which since 2000 have always presented the typical IMRD sections, as revealed by the keyword list of the 2000-2009 corpus sampled in Table 2, where such items as *Background, Methods* and *Conclusions* are the new keywords and which convey the idea of a rigidly-organized pattern like the one traditionally found in scientific research articles. What is noteworthy is the fact that the new keywords characterizing such a corpus seem to have the highest keyness values since they are found at the top of the keyword list. Particularly interesting is the presence of *university* as a keyword revealing the institutionalized framework within which research summarized by posters is carried out. Indeed, all posters of this decade have the clear indication as to the author's position and affiliation, which, if compared with the 1980-1989 sub-corpus (example (9), below), reveals a much clearer expression of institutional identity and authorial credibility (example (8), below):

(8) A. Alonso[1], E. Ferndndez-Jarne[2], C. Fuente[1], R. M. Pajares[1], A. Sanchez Villegas[3], M. A. Martinez-Gonzalez[1]. [1]University of Navarra, Department of Epidemiology and Public Health, Pamplona, Spain; [2]University of Navarra, Department of Cardiology, University Clinic, Pamplona, Spain; [3]University of Las Palmas, Division of Preventive Medicine and Public Health, Las Palmas de Gran Canaria, Spain (P31_2004)

(9) H. Gruchow*, B. Bailey and R. Downer (The Medical College of Wisconsin, Milwaukee, WI 53226) (P1_1983)

In other words, in the 2000s sub-corpus, the stronger the authors' credentials are, the more reliable the methodological approach of the investigation and the better resulting findings are perceived. In addition, the analysis of keywords seems to indicate that a shift in medical perspectives and discourse has occurred, which tends to greater discursive homogeneity in the 2000-2009 corpus than in the other two corpora, due to the generic constraints established with the *Vancouver Style*. The reading of all the posters has allowed the identification of a more narrative style in the 1980-1989 and 1990-1999 corpora and a more promotional style in the 2000-2009 corpus, as the first sentence of each poster excerpted here reveals:

(10) On January 31, 1981, the Centers for Disease Control (CDC) announced that toxic-shock syndrome (TSS) had steadily declined during 9/80-12/80. A review of TSS reporting in Wisconsin was therefore initiated. (P1_1981)

(11) This case-control study was designed to re-evaluate the association of the morphology of breast tissue seen on mammograms with breast cancer risk and to assess the relation of diet, especially intake of fat and vitamin A, to the high-risk mammographic images. (P1_1988)

(12) To evaluate the risk of hospitalization for peptic ulcer disease associated with the use of nonsteroidal anti-inflammatory drugs (NSAIDs), the authors conducted a nested case-control study in the Tennessee Medicaid population, aged 65 years and older, from 1984 through 1986. (P1_1990)

(13) Pancreatic adenocarcinoma is an important cause of death from cancer throughout the world. Aim: To determine the epidemiological characteristics of pancreatic cancer. (P2_1998)

(14) Objectives. To validate a simple questionnaire in Portuguese, for primary migraine diagnosis in epidemiological research. (P2_2001)

(15) Introduction: the prevalence of ECC (Early Childhood Caries) has been a significant public health issue. (P1_2009)

Scientific discourse is the representation of knowledge constructed through the text in terms of problematization, i.e., through the authorial expression of progressive and alternative positions by acknowledging, disclaiming and proclaiming, by means of which the prevailing assumption is presented as needing re-examination, re-conceptualization and/or re-evaluation "in order to provide a ground for the more specific purpose, thesis, point, or argument of an essay" (Barton 1993: 48). In the 1980-1989 and 1990-1999 corpora, problematization, expressed by means of argumentative features, is apparently granted by a much more narrative style than that characterizing the 2000s sub-corpus. We can see, however, that starting from 1998 the syntactic structures used in posters resemble the very syntactic structures exploited by promotional language. In the 2000-2009 sub-corpus problematization is apparently realized through bulleted sentences, a typical feature of advertising (Leech 1966, Goddard 2002, Pennarola 2003), similar to expressions of disjunctive grammar (cf. Leech 1966, Cook 2001), i.e. based on non-finite verbal groups, which can occur both in the *Objective* and in the *Conclusion* sections.

Promotional traits in posters are clearly evident in the use of visuals which are not to be regarded as mere decoration but rather as concept- and theory-laden tools with a greater conceptual role than in oral presentation as they must stand alone without any accompanying explanation and, at the same time, they attract viewers and provide an entry point into conversation (MacIntosh-Murray 2007: 359). Medical discourse within posters, realized by means of theory-laden visuals, seems to be textualized, with short, concise, and accurate wordings (Jaime Sisó 2009) to facilitate the recognition and assimilation of relevant information and is expressed by means of disjunctive forms, which, as we have seen above, resemble the language of advertising, where highly evaluative lexical items are usually found. Evaluation endorses persuasion and mirrors self-promotion. Although I have not come across any particular evident evaluative forms in the 2000-2009 corpus, I nevertheless found the occurrence of the term *socio-* as a keyword. This is particularly interesting, as *socio-* clearly belongs to non-medical domains (cf. Concordance list 2 below) and refers to social aspects such as demography, economics and politics:

Concord

File Edit View Compute Settings Windows Help

N	Concordance	Set	Tag	Word #	t. #	os.	. #	os.	. #	os.	t. #	os.
1	or absence of sealants along with Socio-demographic indicators			4 619	288	4%	0	6%			0	6%
2	or absence of sealants along with Socio-demographic indicators			4 344	277	4%	0	1%			0	1%
3	health programs which considering the socio-economical factors should be			2 543	161	1%	0	7%			0	7%
4	oral health status, unhealthy lifestyles, socio-economic status and ethnic			2 530	160	5%	0	7%			0	7%
5	Children from families with a higher socio-economic status have lower caries			980	59	2%	0	8%			0	8%
6	between caries patterns and socio-economic factors. Results: The			914	55	4%	0	7%			0	7%
7	with molar approximal lesions. The socio-economic status of childrenÕs			888	54	0%	0	7%			0	7%
8	and the relationship between caries and socio-economic factors in pre-school			824	50	2%	0	5%			0	5%
9	P12Ð166 Caries pattern and the related socio-economic factors in pre-school			737	42	3%	0	4%			0	4%
10	study was to get more information on socio-demographic features of HBV			93 393	748	0%	0	7%			0	7%
11	of adolescents and can be related to socio-demographic variables. The			85 896	467	7%	0	9%			0	9%
12	such as obesity is influenced by socio-political and cultural context. This			79 318	212	4%	0	2%			0	2%
13	Domestic violence was associated with socio-demographic and familial variables.			75 369	063	0%	0	8%			0	8%
14	questionnaire comprised questions for socio-demographic characteristics,			75 208	057	9%	0	8%			0	8%
15	pylori infection and its association with socio-demographic factors in a			74 787	039	4%	0	8%			0	8%
16	shift work, socio-demographic and socio-economic indicators, health			65 221	640	6%	0	8%			0	8%
17	about working conditions, shift work, socio-demographic and socio-economic			65 219	640	8%	0	8%			0	8%
18	the population of the HECs on various socio-demographic data of the			63 114	555	7%	0	6%			0	6%
19	found associations were independent of socio-demographic factors, occupational			62 860	545	3%	0	5%			0	5%
20	Western Europe we seem to be beyond socio-economic health differences at			58 528	369	2%	0	1%			0	1%

Concordance list 2. Breakdown of *socio** in the 2000-2009 corpus.

In particular, an extraction of the concordance pattern has revealed that *socio-* mainly occurs in combination with *economic* (177 hits; TTR 0.17), commonly related to social status, social differences, and ethnic background, as we can see in Concordance list 3, below:

Concord

File Edit View Compute Settings Windows Help

N	Concordance	Set	Tag	Word #	t. #	os.	. #	os.	. #	os.	t. #	os.
1	oral health status, unhealthy lifestyles, socio-economic status and ethnic			2 530	160	5%	0	7%			0	7%
2	Children from families with a higher socio-economic status have lower caries			980	59	2%	0	8%			0	8%
3	between caries patterns and socio-economic factors. Results: The			914	55	4%	0	7%			0	7%
4	with molar approximal lesions. The socio-economic status of childrenÕs			888	54	0%	0	7%			0	7%
5	and the relationship between caries and socio-economic factors in pre-school			824	50	2%	0	5%			0	5%
6	P12Ð166 Caries pattern and the related socio-economic factors in pre-school			737	42	3%	0	4%			0	4%
7	shift work, socio-demographic and socio-economic indicators, health			65 221	640	6%	0	8%			0	8%
8	Western Europe we seem to be beyond socio-economic health differences at			58 528	369	2%	0	1%			0	1%
9	HEALTH ECONOMICS Beyond socio-economic health inequalities:			58 276	361	7%	0	1%			0	1%
10	the SDQ (OR 1.04, 95% CI 0.74Ð1.45). Socio-economic status notably			53 927	177	4%	0	6%			0	6%
11	the listed population (demographic and socio-economic) as independent			46 705	895	0%	0	9%			0	9%
12	the quality of care in Argentina. Socio-economic differences in utilization			36 641	475	2%	0	8%			0	8%
13	stroke, osteoarthritis, cancer) and with socio-economic deprivation. Despite the			29 714	185	6%	0	1%			0	1%
14	data (age, gender, education and socio-economic status), medical data			12 328	474	6%	0	3%			0	3%
15	of the boys; adolescents from higher socio-economic groups and who live in			10 140	398	0%	0	1%			0	1%
16	and school children with different socio-economic backgrounds (low-high			1 242	44	8%	0	1%			0	1%
17	P, et al. Inequality in health: socio-economic differentials in mortality			2 553	126	5%	0	3%			0	3%
18	conducted controlling for age, sex, socio-economic status, injury severity			53 478	382	8%	0	8%			0	8%
19	acid use between women who differ in socio-economic ethnic background, and			50 904	262	9%	0	4%			0	4%
20	use between groups with high and low socio-economic status (SES) did not			50 859	260	1%	0	4%			0	4%
21	acid use between women from different socio-economic and ethnic background			50 777	258	3%	0	3%			0	3%
22	to primary health care is related to socio-economic factors. The			50 428	239	4%	0	3%			0	3%

Concordance list 3. Breakdown of *socio-economic* in the 2000-2009 corpus.

There is one occurrence of *socio-political*, indicated in excerpt (16) below:

(16) [...] Policy response to chronic diseases such as obesity is influenced by socio-political and cultural context. (P201_ 2008)

There are also four hits of *socio-cultural*, together with the terms *elements*, *factors*, *differences* and *background*, highlighting social conditions rather than medical implications, as we can see in Concordance list 4:

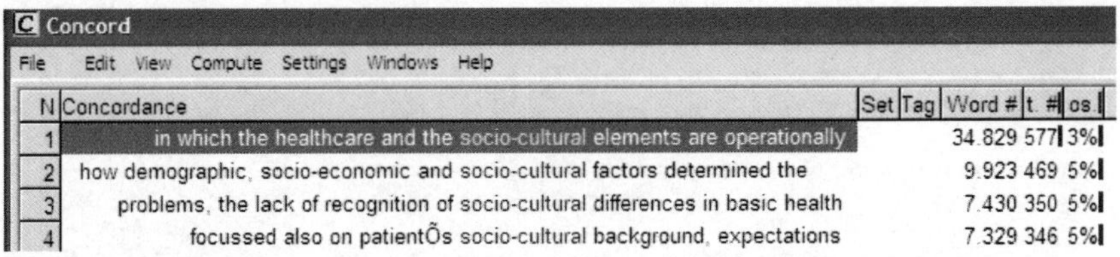

N	Concordance	Set	Tag	Word #	t. #	os.
1	in which the healthcare and the socio-cultural elements are operationally			34.829	577	3%
2	how demographic, socio-economic and socio-cultural factors determined the			9.923	469	5%
3	problems, the lack of recognition of socio-cultural differences in basic health			7.430	350	5%
4	focussed also on patientÕs socio-cultural background, expectations			7.329	346	5%

Concordance list 4. Breakdown of *socio-cultural* in the 2000-2009 corpus.

Since *socio-* mainly concords with *economics*, I checked the frequency list generated for the 2000s sub-corpus and I observed the massive presence of the adjective *economic* (489 hits, TTR 0.49), not found in the previous decades:

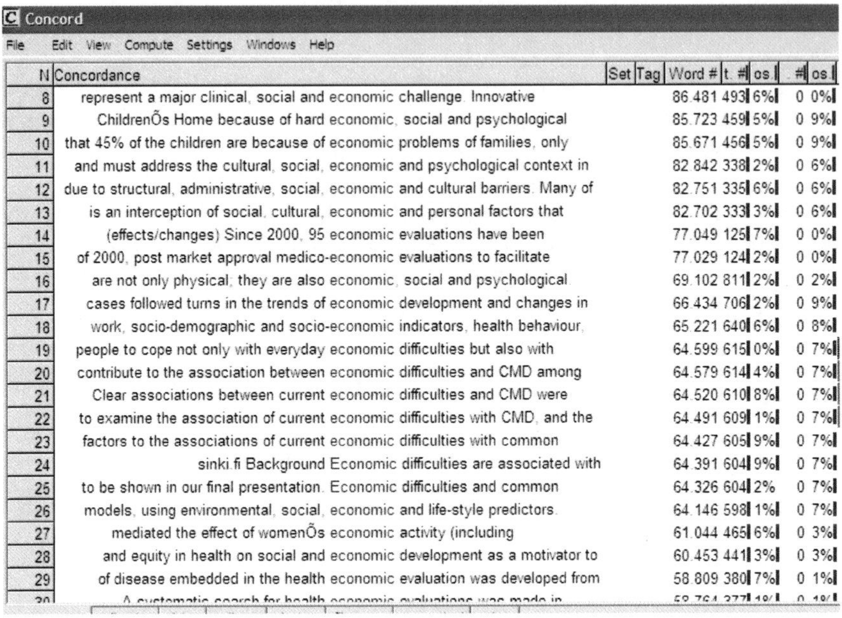

Concord

File Edit View Compute Settings Windows Help

N	Concordance	Set	Tag	Word #	t. #	os.	#	os.
8	represent a major clinical, social and economic challenge. Innovative			86.481	493	6%	0	0%
9	ChildrenÕs Home because of hard economic, social and psychological			85.723	459	5%	0	9%
10	that 45% of the children are because of economic problems of families, only			85.671	456	5%	0	9%
11	and must address the cultural, social, economic and psychological context in			82.842	338	2%	0	6%
12	due to structural, administrative, social, economic and cultural barriers. Many of			82.751	335	6%	0	6%
13	is an interception of social, cultural, economic and personal factors that			82.702	333	3%	0	6%
14	(effects/changes) Since 2000, 95 economic evaluations have been			77.049	125	7%	0	0%
15	of 2000, post market approval medico-economic evaluations to facilitate			77.029	124	2%	0	0%
16	are not only physical; they are also economic, social and psychological.			69.102	811	2%	0	2%
17	cases followed turns in the trends of economic development and changes in			66.434	706	2%	0	9%
18	work, socio-demographic and socio-economic indicators, health behaviour,			65.221	640	6%	0	8%
19	people to cope not only with everyday economic difficulties but also with			64.599	615	0%	0	7%
20	contribute to the association between economic difficulties and CMD among			64.579	614	4%	0	7%
21	Clear associations between current economic difficulties and CMD were			64.520	610	8%	0	7%
22	to examine the association of current economic difficulties with CMD, and the			64.491	609	1%	0	7%
23	factors to the associations of current economic difficulties with common			64.427	605	9%	0	7%
24	sinki.fi Background Economic difficulties are associated with			64.391	604	9%	0	7%
25	to be shown in our final presentation. Economic difficulties and common			64.326	604	2%	0	7%
26	models, using environmental, social, economic and life-style predictors.			64.146	598	1%	0	7%
27	mediated the effect of womenÕs economic activity (including			61.044	465	6%	0	3%
28	and equity in health on social and economic development as a motivator to			60.453	441	3%	0	3%
29	of disease embedded in the health economic evaluation was developed from			58.809	380	7%	0	1%

Concordance list 5. Breakdown of *economic* in the 2000-2009 corpus.

The adjective *economic* collocates in particular with *socio-* (278 hits), as seen above, but also with *status* (70 hits), *differences* (27 hits), *factors* and *inequalities* (21 hits), amongst others, and this seems to have little to do with medicine. A full list of the collocations of the adjective *economic* can be observed in Table 4 below:

L1	*centre*	*L2*	*hits*
socio	*economic*		278
	economic	status	70
	economic	differences	27
	economic	factors	21
	economic	inequalities	21
	economic	conditions	19
	economic	groups	13
	economic	indicator/s	3/13
	economic	position	11
	economic	mortality	9
	economic	strain	8
	economic	evaluation	6
	economic	development	5
	economic	transition	5
	economic	circumstances	5
	economic	deprivation	5
	economic	gradients	4
perceived	*economic*		4
	economic	level	4
	economic	characteristics	3
	economic	analysis	3
political	*economic*		3
	economic	rates	3
	economic	variables	2
	economic	area	1

Table 4. Breakdown of collocates with economic in the 2000-2009 corpus.

A reading of the passages where the collocates of *economic* occur led me to realize that on the one hand, the syntactical construction of poster texts is in line with that of medical discourse traditionally found in research articles, as it is generally characterised by high lexical density (cf. Gotti 2003); on the other, the adjective *economic* seems to assume a meaningful role in the medical discourse of posters, because it is mainly associated with head nouns referring to negative or transitional aspects of society, as excerpt (17) suggests:

(17) The survey revealed difficult socio-economic status as the primary causal factor for low accessibility to health care services. (P53_2002)

What seems to result from the analysis of *economic* is that, within the contexts of medical posters, the medical community acquires a critical role in the discussion regarding aspects related to social problems. In this discussion researchers pose themselves as expert members of the medical social practice subject to global development. Thus, they confirm their active position within society.

Although the adjectives *socio-* and *economic* do not occur consistently in my corpus, they nevertheless seem indicative of a change of perspective in medical discourse: since they do not pertain to the medical field, they probably reflect a profound change in medical society. Their presence apparently confirms what has been pointed out by academic research (Bhatia 2004; Fairclough 2007): not only are global dynamics and universal regulations mediated and translated by the local configurations of resources and ideas which are context-dependent, but such dynamics also give rise to a complex blending of professional cultures when applied to professional practices. The result of these social/professional practices in discursive terms (i.e. the genre) maintains the formality and constraints imposed by the professional community. Yet such a result is represented through a text in which social variations and tensions coexist. In my opinion, the attention posed on social matters within medicine seems to exploit the strategies of prestige advertising, which is used by corporations to emphasize their more prestigious and favourable status than others. In medicine, a field where separation from trade is codified by ethics, pointing at socio-economic, -political and -cultural factors, may seem an indication of self-promotion.

4. Conclusion

Presenting a poster allows authors to more personally interact with the people who are interested in the research presented. Yet, as a poster is viewed together with dozens of others, it has slowly yet consistently become imperative to capture the audience's attention by means of a title, introduction and graphics. In the analysis carried out in this study, by focussing only on the verbal component of posters I tried to identify some significant elements of medical discourse. The diachronic investigation has revealed that:

(i) The language of medical posters has gone from a narrative to a disjunctive mode, the latter resembling bulleted sentences typical of promotional language;

(ii) The presence of the disjunctive mode seems to be necessary to follow the IMRD pattern of scientific discourse, clearly established from the middle of the 1990s;

(iii) There has been a change in the type of lexical items employed in posters:

   a. The keyword lists generated from my three corpora seem to indicate that attention is focused on statistical and empirical analysis in the years 1980-1999 vs. a more precise indication of credentials and a particular attention to *socio-economic, -political* and *-cultural* issues as fundamental factors in health care in the 2000-2009 decade;

   b. The peculiar use of the adjective *economic* in the 2000-2009 corpus suggests an *economic-sensitive* medical community in which the researcher has a role in favouring positive changes.

Although medical posters seem to maintain their integrity when analysed as multimodal communicative events, the diachronic investigation of poster lexico-grammar has revealed a development toward interdiscursive patterns by means of which hybrid generic frameworks are created by appropriating established conventions associated with other genres and professional practices (Bhatia 2007: 393). In medical posters, interdiscursivity results from the application of corporate mentality, globalisation and marketization to scientific discourse. This has required new forms of communicative interaction, i.e., new forms of dialogue representing relations between genres (Fairclough 2005, 2007) by means of which the medical community within the academic world substantiates its active role in the social world by offering strategic plans for social change. Discourse seems therefore endowed with performative power (Fairclough 2007: 10-14) because by describing a reality which needs social action, it realizes that very action in self-promoting forms.

Appendix

Year	*Posters*	*IF*	*Journal Title*	*Issue*	*Pages*
1980	P1-P21	5.454	*American Journal of Epidemiology*	112 (3)	446-452
1981	P1-P20	5.454	*American Journal of Epidemiology*	113 (3)	445-450
1982	P1-P14	5.454	*American Journal of Epidemiology*	116 (3)	580-584
1983	P1-P24	5.454	*American Journal of Epidemiology*	118 (3)	443-449
1984	P1-P27	5.454	*American Journal of Epidemiology*	120 (3)	492-499
1985	P1-P25	5.454	*American Journal of Epidemiology*	122 (3)	540-546
1986	P1-P30	5.454	*American Journal of Epidemiology*	124 (3)	531-539
1987	P1-P44	5.454	*American Journal of Epidemiology*	126 (4)	767-776
1988	P1-P97	5.454	*American Journal of Epidemiology*	128 (4)	918-943
1989	P1-P48	5.454	*American Journal of Epidemiology*	130 (4)	839-852
1990	P1-P113	5.454	*American Journal of Epidemiology*	132 (4)	792-822
1993	P1-P102	5.454	*American Journal of Epidemiology*	138 (8)	644-671
1993-1	P103-P157	3.186	*Journal of Epidemiology & Community Health*	47	412-426
1994	P1-P66	3.186	*Journal of Epidemiology & Community Health*	48	509-525
1995	P1-P 3	3.186	*Journal of Epidemiology & Community Health*	49	S78-S80
1998	P1-P 172	1.77	*Digestion*	59 (3)	213-265
2001	P1-P 2	3.68	*Cephalalgia*	21 (4)	303-312
2002	P1-P 242	2.76	*European Journal of Public Health*	12 (1)	35-110
2003	P1-P 151	2.76	*European Journal of Public Health*	13(2)	91-136
2004	P1-P 209	3.186	*Journal of Epidemiology & Community Health*	58 (suppl.1)	A64-A125
2005	P1-P 122	4.93	*Critical Care*	9 (suppl. 2)	S1-S97

Year	*Posters*	*IF*	*Journal Title*	*Issue*	*Pages*
2006	P1-P 424	4.93	*Critical Care*	10 (suppl. 1)	S1-S194
2007	P1-P 12	1.07	*Journal of Intervention Cardiac Electrophysiology*	18	99-105
2007	P13-P292	2.76	*Journal of Epidemiology & Public Health*	17	136-240
2008	P1-P 233	2.76	*Journal of Epidemiology & Public Health*	18 (1)	130
2009	P1-P 215	1.07	*International Journal of Pædiatric Dentistry*	19 (suppl.1)	66-170

MARTIN SOLLY

Dialogic Monologues: Commencement Speeches as an Evolving Genre*

1. Introduction

The historical evolution of genres is currently considered a fruitful area of applied linguistics research, enabling researchers to examine the developing practices and changing textual dynamics of professional and academic genres diachronically (Bhatia 2004, Berkenkotter 2008a). In this chapter the concept of genre evolution is applied to commencement speeches (CS), those speeches typically delivered at graduation ceremonies in North American universities by prominent professional and academic figures. Elsewhere (Solly 2011) I have outlined some of the main discourse features of CS, identifying them as socially situated and socially ratified communicative events (Fairclough 2003; Scollon 2008), which take place in a defined higher education context, which share a set of communicative purposes (Swales 1990), and which can thus be considered as a distinct but dynamic genre (Berkenkotter/Huckin 1995). In this study I take a diachronic perspective and, starting with George C. Marshall's famous 1947 Harvard CS where his plan to provide economic aid for post-war Europe was unveiled for the first time, focus on a small sample of CS delivered over the years till 2010, in order to see how the genre has evolved.

In particular this chapter examines the rhetorical strategies and moves (Biber *et al.* 2007) deployed by CS speakers, emphasising the increasingly interactive nature of CS and suggesting that it is in large

* This chapter is published under the auspices of the Italian national research project: 'Tension and change in English domain-specific genres' funded by the Italian Ministry of University (prot. 2007JCY9Y9).

part determined by two main factors which have been changing over the years: the pervasive use of multimodality in an increasingly global context (Machin/van Leeuwen 2007) and the need for the speakers to engage actively with their audiences.

2. Context and the framing of a genre

Recent work on genre theory has also highlighted a significant shift of emphasis from text to context (Bhatia 2008a: 166) and genre researchers have been advised to use ethnographic methods "to focus on discovering the activities, attitudes, beliefs, values and patterns of behaviour of the discourse community engaging in the genre or genres which is/are the focus of study" (Flowerdew/Wan 2010: 81). While this chapter does not analyse CS from an ethnographic point of view, it does appreciate the importance underlined by the authors of seeing the genre as embedded in an increasingly wide-ranging discourse community. Indeed, following Giltrow and Stein's contention that the notion of context can range from context defined by "the narrowest co-presence heuristics operative in focused interaction" to context defined by "wider historical measures or by the broadest critical-theoretical analyses" (2009: 2), I would suggest that the context in which CS are delivered has two main dimensions: first, the restricted context of the audience actually present; second the wider audience of the world at large. While the first of these has hardly changed in size over the last fifty years, the second has expanded exponentially since the advent of the Internet. These days CS are usually filmed and swiftly made available online in both transcript and podcast versions on the host university's website. They can thus be easily accessed by a far larger and potentially much more varied audience as they are no longer limited to those actually present at the graduation ceremony or to the press reporting of the event.

Moreover, while CS certainly provide speakers with the possibility of inscribing their speeches into the local history, traditions and folklore of the academic institution in which they are delivered, they also provide speakers with the possibility of using their speech to occupy a potentially far more significant niche in the larger, much wider scale history of their nation and perhaps even that of the world.

CS fit into the broad category of very formal academic lectures and, as such, provide speakers with a rather special forum. One where speakers are guaranteed the presence of a large distinguished audience and probable press coverage, but at the same time one where they are also guaranteed immunity from interruption and where there will be no questions at the end. Thus they are occasions where speakers can launch ideas and initiatives without being immediately challenged and where they can also take advantage of the academic *gravitas* provided by the prestigious occasion and the university setting. US Secretary of State George C. Marshall chose the CS he delivered at Harvard on 5 June 1947 as the platform to present his famous plan to provide economic aid for post-war Europe, rather than unveil it directly to Congress or the recently formed United Nations.

> Addressing a lay crowd gave Marshall more leverage. Strict isolationists in Congress might have dismissed the recovery strategy immediately, forcing him into an uphill battle. Instead, the largely conservative Harvard audience and the media in attendance applauded the plan, paving the way for its reception on Capitol Hill. (Conger 2009)

Here is a recent Italian example. At the time of writing, Mario Draghi, Governor of the Bank of Italy, had just used the *lectio magistralis* he was delivering at the University of Ancona (5 November 2010) to make a frank analysis of the current state of the Italian economy and at the same time present a radical request for employers in Italy to employ young people with permanent rather than temporary contracts. His speech was immediately reported by all the major Italian newspapers and radio and television news programmes, creating a certain *furore*. His choice of the formal academic setting as the podium from which to launch his potentially controversial plea seems very similar to George C. Marshall's choice of Harvard University to launch what became known as the Marshall Plan.

CS have a long tradition and have become an integral part of the North American university's academic year. Indeed the CS tradition is emphasised by the universities themselves. Harvard University's Commencement Office for example notes that its commencement traditions date back to 1642. And also by the speakers: Barack Obama points out to the Notre Dame University graduates on 17 May 2009 that "One hundred and sixty three classes of Notre Dame graduates have sat where you are today". Certainly the format, setting and academic dress have remained frozen in time. Ursula Le Guin, in a powerful CS delivered at Mills College (CA) on 19 May 1983 on what she described as a "rare chance: to speak aloud in public in the language of women", observed that:

(1) [...] commencements are usually operated under the unspoken agreement that everybody graduating is either male or ought to be. That's why we are all wearing these twelfth-century dresses that look so great on men and make women look either like a mushroom or a pregnant stork. Intellectual tradition is male. (Ursula Le Guin, Mills College, 19 May 1983)

Today the same academic dress is still worn, with graduates and speakers alike still immortalised proudly wearing the medieval apparel. This wearing of the traditional academic dress forms part of the ancient and medieval use of liveries, uniforms, flags, badges, coats of arms and so on to represent social identities and membership, the belonging to a group. But the academic dress also carries a legal valence. The right to wear particular academic robes (cap and gown) pertains only to holders of a specific degree awarded by a specific academic institution, or by members of that institution who hold a certain rank. This is still the case and the legal implications are closely linked to the use of brand logos and trademarks which have a strong communicative and commercial as well as legal significance. At a commercial level trademarks act as a sign of consistent quality, enabling businesses to attract and retain customers, and to promote corporate identity (Solly 2002), a useful quality in the age of the entrepreneurial university. The wearing of robes at academic ceremonies is therefore a strongly communicative sign, establishing and transmitting the university's tradition and power, and promoting its brand image as well as its *de facto* right to award degrees.

If the CS speakers have been awarded an honorary degree (very often the case), they will wear the robes which holders of the degree at that university are entitled to wear. But not always. On 8 June 2001 NASA Administrator Daniel Goldin delivered the CS at the Massachusetts Institute of Technology. During his speech he (excerpt 2) specifically refers to the academic regalia he is wearing. He uses the reference to the robes to link his speech to the University of Padua and the fact that Galileo Galilei would have worn the same robes at his lectures. This enables him to introduce the theme of his CS (on Galileo, scientific discoveries and truth), but at the same time the wearing of the University of Padua's academic regalia frames the speech in traditions far more ancient than MIT's. And, as Goldin points out, he has primarily chosen to provide the graduating students with a striking visual image in the hope that they will remember it and, above all, that they will remember Galileo's lessons and life.

(2) I would like to start today by telling you a little about the robe I'm wearing. Exactly one year ago today, I was invited to give a lecture at the University of Padua, in Italy, after receiving an honorary doctorate.
The University of Padua is where one of my heroes, a professor by the name of Galileo Galilei, built his first telescope. [Where] Galileo delivered countless lectures to some of the best and brightest students in the world. […]
The robe I'm wearing today is what I wore at Padua that day. It is the garb that lecturers have worn at Padua since Galileo's time. Actually, they had to wear this robe and this cap.
Contrary to what many of you are thinking, I'm not wearing this because I lost some cruel bet. And I'm not wearing it because I'm the victim of the latest MIT hack. At least not yet.
I wear this cap and gown because I want to give you a visual image. As you lift off from this wonderful institution to explore and discover whatever mysteries life may hold. You may not remember me. But perhaps you will remember the cap and gown.
More than that, I hope that each and every one of you remembers Galileo. Not necessarily his lectures, but his lessons and his life. (Daniel Goldin, MIT, 8 June 2001, my emphasis)

Thus the framing of CS remains highly traditional, conforming to the traditional packaging and format of the host institution.

However, the CS genre has not remained static. What has changed are the language and style of the delivery and the manner of its diffusion (the latter reflecting the multimodality made possible by the advent of the new technologies). These days universities will make certain that their name and logo are extremely visible, also on the web pages where the CS transcripts or podcasts are posted. When philosopher Martha Nussbaum delivered the 189th commencement at Colgate University on 16 May 2010, she was standing (as we can see from the university webpage where the CS is available: <http://blogs.colgate.edu/2010/05/s.html>) at the podium bearing the name of the university in large letters, wearing the university's black and red robes; the red matching the red of the university's white and red logo which is projected vast upon the screen behind the speaker. This is clear evidence of how the exploitation of the promotional possibilities provided by the visual framing of the CS has evolved over the years.

3. Performance and persuasion

CS have become an increasingly public moment in the academic year and are often widely reported in the press as well as being available on the Internet. The CS delivered by Steve Jobs at Stanford University on 14 June 2005 and that by David Foster Wallace at Kenyon College (OH) on 21 May 2005 have achieved cult status on the North American campus circuit (and beyond) precisely because they are available as podcasts: they can therefore be enjoyed privately and intimately by a vast and very different audience from that of George C. Marshall's 5 June 1947 CS. But their success is also due to the persuasive and dialogical rhetorical strategies with which the speakers engage their audiences. Thus while the traditions remain frozen in the setting and framing of CS, the discursive innovations reflect the diachronic changes in the genre as well as the personal choices taken by the speakers themselves.

CS speakers are aware of the pressures to conform to the genre and they generally do so, albeit deploying a whole range of discursive strategies in order to interact personally with their audience, thus each CS also bears the hallmark identity of its deliverer while at the same time playing an important part in the construction/representation of that identity (Wodak 2009). An identity which is encapsulated in the discourse structure and rhetorical organisation of the CS, in its use of narrative, story-telling, anecdotes, exemplification, humour, references to shared experience and so on, as well as the specific lexico-grammatical and stylistic choices, such as personal pronouns, imperatives, questions, informal language, indirect speech and repetition in its various forms. The study of how these language features are deployed, for example seeing the differences between spoken and written language as "scales along which individual texts can be plotted" (McCarthy 2001: 93), is a useful way of charting some of the diachronic developments in the language of CS.

Although delivered by one speaker many CS are likely to be the work of several hands, ranging from professional speechwriters to friends or colleagues invited to take a look at draft versions of the CS and give an opinion before its presentation. Nevertheless, irrespective of whether the speakers have actually written all their CS themselves, CS are delivered by the speakers and are considered to be theirs. As Wodak (2009: 2) points out, political speeches are "usually written by 'spin doctors', but *performed* by the politicians themselves" (my emphasis). Thus the ownership of the text and its style belong to the speech deliverer and are taken to be so by the audience. The notion of 'performance' is interesting in the analysis of CS and is "necessarily and intrinsically related to the metaphor of 'being in the theatre and on stage'" (Wodak 2009: 9). The 'performance' of a CS is staged by the host institution and framed in the academic setting as we have seen, and thus the ownership of the 'performance' is also shared by the host institution whose robes the CS speakers are likely to be wearing.

As noted earlier, CS speakers are responsible both for the content of their CS and also for the style of their presentation which is often extremely personal, with factors such as appearance, manner and charisma strongly affecting the dramatic realisation of their performance and impacting on its appeal and credibility, in short its persuasive force. This personalisation is actually typical of much academic lecturing and indeed of academic communication in general which, as Gotti observes, "[...] remains in many ways a highly personal affair" (2009: 9).

One of the most important factors underpinning the persuasive force of speeches is the way the speakers negotiate their identities with their audience, in other words how they "go about signalling and recognising the roles and discursive positions which they consider themselves and their interlocutors to be occupying in a specific situation" (Riley 2006: 297). Riley, like a number of other applied linguists, (see, for example, Biber *et al.* 2007), finds the Aristotelian concept of *ethos* particularly useful in the analysis of negotiated identity. *Ethos* enables speakers to construct and convey a trustworthy, credible image of themselves (or the identity they have chosen to adopt and project) to their hearers. It makes use of various rhetorical devices to enhance the speaker's positive, reliable character and sound judgment, such as the narration of their own first-hand experience, the demonstration of their respect for the audience's interests and opinions, the emphasis on shared interests, experience and points of view. For Riley,

> Foregrounding a specific aspect of one's social identity in order to orient one's interlocutors' expectations and perceptions is a major aspect of the construction of ethos. By laying claims to membership of a particular social category, speakers aim to establish their competence (knowledge, experience) in the relevant domain, thereby confirming their credibility. (2006: 308)

As regards the construction, recontextualisation and performance of identities, Wodak stresses the importance of the 'negotiation of presuppositions' as a significant indicator of 'shared knowledge' or of inclusion/exclusion from knowledge: those who 'know' also share the same assumptions, meanings and presuppositions (2009: 26). By careful use of presuppositions, insinuations and implicatures speakers use persuasive rhetoric to construct intertextuality and interactivity, making their monologues communicative, dialogic. At the same time they can also build solidarity through humour, orienting their listeners with allusions to shared experiences, and careful choice of evaluative language.

In their study of moves in fundraising letters Biber *et al.* (2007) suggest that analysis of rhetorical appeals reveals the persuasive roles of text sections as well as their communicative and informative functions. For them each move has a local purpose, but also contributes to the overall rhetorical purpose of a text. Thus, as well as describing texts according to their semantic/functional purpose, as in conventional genre move analysis, they highlight the importance of the Aristotelian categories of *ethos*, *logos* and *pathos* which "can [also] be identified in texts using a top-down, discourse-structural process" (Biber *et al.* 2007: 122). The combination of move analysis and rhetorical persuasion analysis seems particularly useful for the purposes of the present investigation of CS, in that it enables the researcher to identify and analyse the rhetorical strategies deployed by speakers as well as the genre organisation and structuring. A top-down qualitative methodological approach has therefore been employed to examine the purpose and characteristics of the persuasive roles of text sections.

4. Opening moves and gambits

In this section I look more closely at the opening moves of CS. The opening move generally includes two steps: greeting those present and thanking the institution (Solly 2011). Both of these are highly interactional – immediately engaging the host institution (often with the naming of specific members) and the graduating students. But as we will see the opening move is sometimes very different and can be quite complex.

Steve Jobs in excerpt (3) greets those present, thanks the institution (saying how 'honored' he is to be present), compliments it as being "one of the finest universities in the world", and then also refers to his own university (non)career:

(3) I am honored to be with you today at your commencement from one of the finest universities in the world. I never graduated from college. Truth be told, this is the closest I've ever gotten to a college graduation. Today I want to tell you three stories from my life. That's it. No big deal. Just three stories. (Steve Jobs, Stanford University, 14 June 2005)

This opening is extremely interactional with its use of personal pronouns and the swift change from formal to conversational mode. It also contains another aspect often deployed in the opening paragraphs of CS, which, taking my cue from the language of chess, I will define as an 'opening gambit'. In chess a gambit is a chess opening in which a player risks losing one or more minor pieces in order to gain a positional advantage. A gambit is also defined as "an opening move in a discussion etc" (OED). Here it seems appropriate to use the term 'gambit'[1] alongside the classical genre analysis terms of 'move' and 'step', as it enables me to emphasise the rhetorical strategy implicit within the move. And also because it implies a fundamentally dialogic relationship, between the two players in a game of chess, between the two parties in a discussion; here in the apparently monologic CS (only one person is actually speaking). In the case of excerpt (3) I would contend that Jobs is introducing a self-deprecatory even negative aspect of his own career "this is the closest I've ever gotten to [...] graduation", precisely to gain a positional advantage over his audience.

In excerpts (4) and (5) the two speakers use the same opening gambit. They both admit to feeling fear about the delivery of their speeches in order to gain a potential advantage, in this case I would suggest that they risk displaying their own human frailty and fear in order to position themselves as highly credible, reliable human beings. In excerpt (4), George C. Marshall opens his 1947 CS at Harvard University:

(4) I'm profoundly grateful and touched by the great distinction and honor and great compliment accorded me by the authorities of Harvard this morning. *I'm overwhelmed, as a matter of fact, and I'm rather fearful of my inability to maintain such a rating as you've been generous enough to accord to me.* In these historic and lovely surroundings, this perfect day, and this very wonderful assembly, it is a tremendously impressive thing to an individual in my position. But to speak more seriously, I need not tell you, gentlemen, that the world situation is very serious. (George C. Marshall, Harvard University, 5 June 1947, my emphasis)

In excerpt (5) British writer J. K. Rowling is also at Harvard University, addressing the graduating students in 2008:

---

1 The term 'gambit' has also been used by linguists to signify conversational strategy signals in the area of discourse markers (see for example Keller 1979, Müller 2005: 22); here I use the term differently.

(5) President Faust, members of the Harvard Corporation and the Board of Overseers, members of the faculty, proud parents, and, above all, graduates. The first thing I would like to say is 'thank you.' Not only has Harvard given me an extraordinary honour, *but the weeks of fear and nausea I have endured at the thought of giving this commencement address have made me lose weight.* A win-win situation! Now all I have to do is take deep breaths, squint at the red banners and convince myself that I am at the world's largest Gryffindor reunion. (J. K. Rowling, Harvard University, 5 June 2008, my emphasis)

In excerpt (6) however, Martha Nussbaum's 2010 CS opening is more similar to George Marshall's (excerpt 4), although without the reference to her fear. Again the hosting institution is thanked and the celebratory joy of the occasion referred to. In a similar opening to excerpt (5) the university president, faculty members, family members and graduating students are all specifically mentioned. But then, as in excerpt (4), the speaker moves straight to the gravity and seriousness of the world economic situation and, in this case, the threat to liberal education.

(6) Interim President Roelofs, members of the faculty, trustees, alumni, family members and other guests, and above all, graduates. On this joyful day, we are here to celebrate a wonderful group of young people who have achieved so much, graduating from one of the premier liberal arts universities in the United States, and who face exciting prospects for the future.
The type of liberal education you have received, however, is under assault all over the world in our time of economic anxiety, as all nations compete to keep or increase their share in the global market. (Martha Nussbaum, Colgate University, 16 May 2010)

5. Stories, anecdotes and phatic bonding

In the final sentence of his opening paragraph (excerpt 3) Steve Jobs says that he will tell his audience 'three stories' and the three stories he then narrates provide the basis of his highly successful CS. The use of narrative and the telling of stories are an accepted, or rather expected, part of CS; indeed they have become a 'standard requirement' of the CS genre (David Foster Wallace, Kenyon College, 2005). The stories told are often anecdotal (in the case of Steve Jobs the anecdotes describe episodes from his own life), but not necessarily so. Sometimes they recount stories from history or literature; thus they are often first-hand but they can also be second-hand, related through indirect speech. The recounting of stories serves a variety of purposes and Riley (2006: 311) observes that anecdotes have a wide range of functions, including the exemplification, argumentation, explanation, and justification of knowledge, beliefs, values and attitudes. He also notes that anecdotes have an important interactive role, enabling groups of friends for example to affirm and/or reaffirm their shared social values, their group identity, through phatic bonding.

> This phatic bonding is clearly part of the ritual of constructing or maintaining *group* identity, and as such presents a number of specific discourse characteristics, such as high levels of co-construction and appeals to shared memories. (Riley 2006: 311, original emphasis)

I would suggest that the concept of phatic bonding is useful in the analysis of the rhetorical strategies deployed in CS. After all CS represent an important moment in the relationship between the various groups actually present: the graduating students, their parents, the institutional representatives, the CS speaker. The phatic bonding is primarily for them, not for the wider audience of the world at large.

Phatic bonding is also achieved through the use of shared knowledge and codes which enable the speakers to create complicity with their audience, aligning themselves with their audience. In George W. Bush's 12 December 2008 CS to Texas A&M University he constantly refers to the university's traditions and some parts of the President's discourse could be difficult to follow without a glossary as can be seen for example in excerpt (7). But the solemn rituals demonstrate the strength of the bonds among the graduating student audience, as Bush emphasises, and also enable the speaker, by referring so intimately and often to them, to share in and empathise with those bonds:

(7) This campus is home to solemn rituals that demonstrate the strength of your bonds. In playing of *Silver Taps* to honor fallen classmates, in the reunion of students and alumni to read the roll call at *Muster*, and in wearing of your

> *timeless rings*, you affirm a powerful truth: Once an *Aggie*, always an *Aggie*. (George W. Bush, Texas A&M University, 12 December 2008, my emphasis)

Anecdotes also enable the speakers to construct the image of themselves that they are presenting to the wider world:

> Taken together, then, it can be seen that anecdotes are vehicles for the encapsulation of remembered experience, and as such are the fundamental materials from which selves are constituted and self-expression is made. It is not surprising, then, to find that they are very often instruments for self presentation and the construction of ethos. (Riley 2006: 311-312)

I would also suggest that it is important not to decontextualize the anecdotes and stories from the CS setting. De Fina (2009: 237) echoes Schegloff's criticism of Labov's model of narrative structure precisely because of this decontextualisation, and reiterates the need for an interactional approach to narrative genres. Referring to the literature on narrative, she points out

> [...] that many narratives are co-constructed, co-evaluated, that they are modified through interaction with audiences, etc. in sum, that they escape the neat organization of monological tellings. (De Fina 2009: 238)

In short, many narratives are dialogic.

A different kind of opening is that chosen by Martha Nussbaum in the CS she delivered on 16 May 2003 to the graduating students at Georgetown University (excerpt 8).

(8) President DeGioia, faculty, parents and friends, and, especially, graduates: On this day of celebration, I want to ask you to pause for a minute, and to think of the ending of a tragic drama, Euripides' The Trojan Women. The towers of Troy are burning. All that is left of the once-proud city is a group of ragged women, bound for slavery, their husbands dead in battle, their sons murdered by the conquering Greeks, their daughters raped. (Martha C. Nussbaum, Georgetown University, 16 May 2003)

After greeting the institution, the graduating students and their families and friends, she asks her audience to recall the ending of a famous play by Euripedes. The audience might have a moment of panic as they dredge their memory to remember the play (if indeed they have actually read or seen it). But there is no need to worry as Nussbaum then recounts the story, using the narrative to create a powerful mental image of the towers of Troy burning and the destruction of the city. The opening move is highly successful. In this case the CS speaker has positioned herself as the expert on Ancient Greek literature and she then goes on to parallel the aftermath of the Fall of Troy with the aftermath of the destruction of the Twin Towers in New York on 11 September 2001, focusing on the twin themes of her talk: compassion and global responsibility.

As regards narrative, in his 1947 CS George C. Marshall illustrates his presentation of the post-war economic plight in Europe by telling the story of the typical farmer and the typical city dweller, a kind of postwar variation on the classic country mouse, city mouse fable to illustrate the problem of the division of labor and how it is threatened with breakdown:

(9) There is a phase of this matter which is both interesting and serious. The farmer has always produced the foodstuffs to exchange with the city dweller for the other necessities of life. This division of labor is the basis of modern civilization. At the present time it is threatened with breakdown. The town and city industries are not producing adequate goods to exchange with the food producing farmer. Raw materials and fuel are in short supply. Machinery is lacking or worn out. The farmer or the peasant cannot find the goods for sale which *he* desires to purchase. So the sale of *his* farm produce for money which *he* cannot use seems to *him* an unprofitable transaction. *He*, therefore, has withdrawn many fields from crop cultivation and is using them for grazing. *He* feeds more grain to stock and finds for *himself* and *his* family an ample supply of food, however short *he* may be on clothing and the other ordinary gadgets of civilization. Meanwhile people in the cities are short of food and fuel. So the governments are forced to use their foreign money and credits to procure these necessities abroad. This process exhausts funds which are urgently needed for reconstruction. Thus a very serious situation is rapidly developing which bodes no good for the world. The modern system of the division of labor upon which the exchange of products is based is in danger of breaking down. (George C. Marshall, Harvard University, 5 June 1947, my emphasis)

Marshall presents the story with great simplicity, using the narrative as a vehicle for exemplifying the disastrous situation in Europe and at the same time for explaining in simple terms the complicated economic picture. He knows that such a story will resonate with many voters in post-war (but also post-depression) America. Interestingly he feels the need to apologise at the end of his CS for his use of "rather technical discussion".

(10) I am sorry that on each occasion I have said something publicly in regard to our international situation, I've been forced by the necessities of the case to enter into rather technical discussion. (George C. Marshall, Harvard University, 5 June 1947)

In actual fact Marshall has no real need to apologise: the use of clear language and the linking of the economic concepts through narrative and images to ones his audience can relate to easily, alongside his personalization of the farmer/peasant with the use of the personal pronouns and adjectives (*he*, *him*, *his*, *himself*) guarantee the communicative success of his speech. But by apologizing he contributes again to the *human* aspect of his character he has introduced in excerpt (4).

Humour is also used by speakers to engage with their audiences and its success can be measured by the indication of laughter in some of the transcripts or by the actual laughter and clapping heard in the podcasts. In the CS openings presented here the humour is evident in some of the excerpts. In excerpt (11) below, for example, Daniel Goldin (see excerpt 2) uses the reference to the robes he is wearing in order to josh the students over some of their possible pastimes and pranks (betting and hacking):

(11) Contrary to what many of you are thinking, I'm not wearing this because I lost some cruel *bet*. And I'm not wearing it because I'm the victim of the latest MIT *hack*. At least not yet. (Daniel Goldin, MIT, 8 June 2001, my emphasis)

In excerpt (5) above J. K. Rowling jokes about that fact that her pre-CS fear and nausea have enabled her to lose weight, a "win-win situation", and the red banners of Harvard are the cue to her humorously linking the setting and the students present to Gryffindor, one of the four school houses of Hogwarts School of Witchcraft and Wizardry, the very traditional school featured in her Harry Potter books. But although the humour and laughter are part of the entertainment factor increasingly expected by CS audiences, they are not always present.

There is no humour in George C. Marshall's 1947 CS, nor is there any in Martha Nussbaum's 2003 and 2010 CS.

6. Conclusion

Without doubt developments in multimodal communication have had a significant impact on the changes that have taken place in the CS genre. The genre evolution has also been strongly influenced by the marketisation of higher education and by social change, as can be seen in the increasingly conversational style chosen by so many CS speakers. It is also clear that the monologues are becoming increasingly dialogic, with speakers actively engaging their audiences and using a wide range of interactive language strategies, including: directly addressing the audience (you); use of imperatives; use of questions; use of humour; use of narrative; use of shared codes, experience and ideals. Despite this, some of the main characteristics of the CS genre have not changed at all over the years, for example, their traditional framing in the academic setting which has remained more or less frozen in time.

PAULA DE SANTIAGO

Diachronic Evolution of Scientific Popularised Articles in Online Newspapers: Critical Reflections on *El Mundo*

1. Introduction

The American newspaper industry has been experimenting since the seventies with informative genres of new creation to provide information about health and science. These genres became most successful when they started to appear in specialised sections of online newspapers. The subsequent evolution of web genres has led to a combination of the traditional characteristics of print and the advantages of electronic media. This combination, however, has taken a variety of forms. To account for them, it is necessary to take into consideration the degrees of dynamism which have been indentified within online media (López *et al.* 2005). The concept of dynamism refers to the degree of use that the new media make of the potentials offered by the online format, which can be summed up under the headlines of constant updating, multimodality, hypertextuality and interactivity. According to López *et al.* (2005: 45), "the greater the degree of adaptation achieved by each of the four indicators, the greater the resulting degree of dynamism of the cybermedium".

This chapter reports on a particularly important case, i.e *WebSalud*, which the Spanish newspaper *El Mundo* publishes in collaboration with *Amgen*, a leading biotechnology company. Before starting with the analysis, an introductory overview will be given of three relevant aspects which have played an especially important role in the evolution of mass media, including online newspapers and, therefore, *El Mundo*. These aspects concern 1) the origin of popularised articles; 2) the introduction of science into the specialised sections of newspapers; and 3) the role of the Internet in these developments.

1.1. New Journalism

The origin of popularised articles is to be found in the *informative genres of new creation*, which were developed in the seventies in the United States to inform citizens about the revolutionary events which were taking place at the time. War protests, the hippie movement and the rise of other counter-culture youth identities provided all the right ingredients needed to drive what came to be known as New Journalism. These stories and events did not fit into the traditional hard-news format of mainstream journalism. Whether they loved it or hated it, journalists and media critics of the time acknowledged the changing ideals of acceptable journalism. This journalism of new creation (also called literary journalism) cannot be considered a school, but rather indicates a form of news discourse featuring a tendency towards a mixture of conventional journalism, tale and short story (Bernal/Chillón 1985: 38). It broke the moulds of conventional journalistic genres because it showed the writer's subjectivity, deploying highly innovative linguistic resources, especially rhetorical devices which used to be avoided in traditional journalism. New journalists differed from the conventional ones because they went beyond just reporting the facts in a traditional, dry, inverted-pyramid structure; rather, they saw themselves as culture critics (Duwe 2005: 5). They considered inclusion of observation and narrative description essential to providing an accurate account of an event, topic or issue.

1.2. The introduction of science in specialised sections of newspapers

During the 20th century, scientific knowledge gradually became more and more widespread. Many scholars (among them Morales 2002: XV; Rensberger 2009: 2) agree on tracing the origin of this development back to the converging interest of scientific societies and press agencies that occurred during World War I and in the period between World War I and World War II. US science reporters were the first ones to decide that the time had come for scientists to take them more seriously. To this purpose, in 1934 they founded the National Association of Science Writers (NASW). The importance of this association has persisted to this day, as the principles in its mission are still central to popular science writing in the Digital age: "NASW members would talk of how to make science journalism 'more effective', by which they meant making it result in the wider public's appreciation of science" (Rensberger 2009: 3). By the 1970s, there was no way science journalists could ignore the social and political implications of their topic. Science journalism boomed, starting with the birth of a science supplement in *The New York Times* in 1978: the *Science Times*. Spanish newspapers followed this model in the late 1980s: *La Vanguardia* started publishing *Ciencia y Tecnología*, *El País* launched *Futuro and Salud*, and *El Mundo* began issuing *Salud.*

1.3. The Internet

In the evolution of mass media, there are two accelerating factors related to the Internet. They are the speed of technological progress and the increased dependency, on the part of the public at large, on information and communication. When web genres having a print antecedent are placed in the new, hypertextual and multimodal medium of the Internet, it is common for them to acquire new properties (Askehave/Ellerup Nielsen 2004: 11). Although Spanish newspapers have felt the need to adjust to new technologies, their approach seems to be to replicate fairly closely the content and form of printed newspapers. Many of them do not take advantage of the functional possibilities presented by the Web. Indeed, some newspapers are still publishing printed health and science supplements online without any changes. This is why in most cases online editions of newspapers have not benefited from the full range of affordances provided by the Internet technology. However, the idea that printed popularised articles must remain in text form only, perhaps accompanied by an occasional still photograph, has become outdated on the Internet (Gunter 2003: viii). The increased use of computer mediated technology to present information has meant that journalists can no longer rely purely on traditional reporting skills. We believe that if science journalists are to regain relevance in society not only must they learn enough science to analyse and interpret the findings, but they must also master the new medium (Askhehave/Ellerup Nielsen 2004: 10).

2. The case of *El Mundo*

The purpose of this section is to account for and describe the evolution of web genres in *El Mundo*. To this end, a comparison between the online supplement *Salud* and the *WebSalud* of the above-mentioned newspaper will be carried out. Both sections include scientific popularised articles (SPAs) and both are published online. However, while *Salud* SPAs are written originally for the printed supplement and then transferred to the Internet in a very similar format as that of their printed counterparts, the ones from *WebSalud* do not always materialise in a printed version. Furthermore, *WebSalud* SPAs are of a much more dynamic nature; they can be changed or replaced within hours.

The comparison will first focus on some examples of the SPAs extracted from *Salud* and *WebSalud*, with attention being paid to their form and functionality. At a later stage a more subjective perspective will be adopted, and the results of the textual analysis will be compared with the insights provided by three interviews conducted with the Health Department of the newspaper *El Mundo* in December 2009. Based on the professional perspective of the Department's members, an attempt will be made to account for the rapid growth of *WebSalud* SPAs. In this respect, Angeles López, Health Department coordinator of *El Mundo*, has been very helpful.

2.1. Salud and WebSalud

This paragraph is based on close reading of several articles from both *Salud* and *WebSalud*. However, for the sake of simplicity, the discussion will make reference to one single representative example from each of the two supplements, in order to highlight the most important differences between them. The articles selected for analysis are 'La células madre neuronales de adultos son pluripotenciales' (Jauregui 2000) from *Salud*:

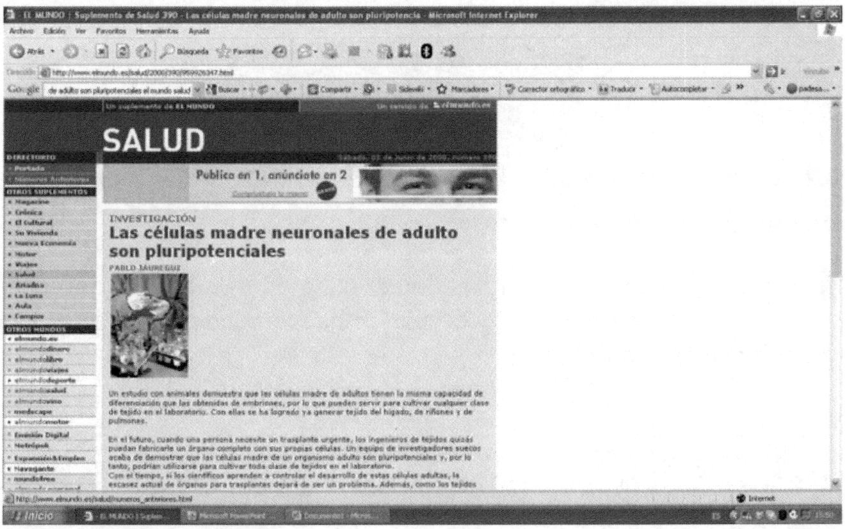

Figure 1. Screenshot of *Salud*.

and 'La reprogramación celular, estrella de la investigación en 2008' (López 2008a) from *WebSalud*:

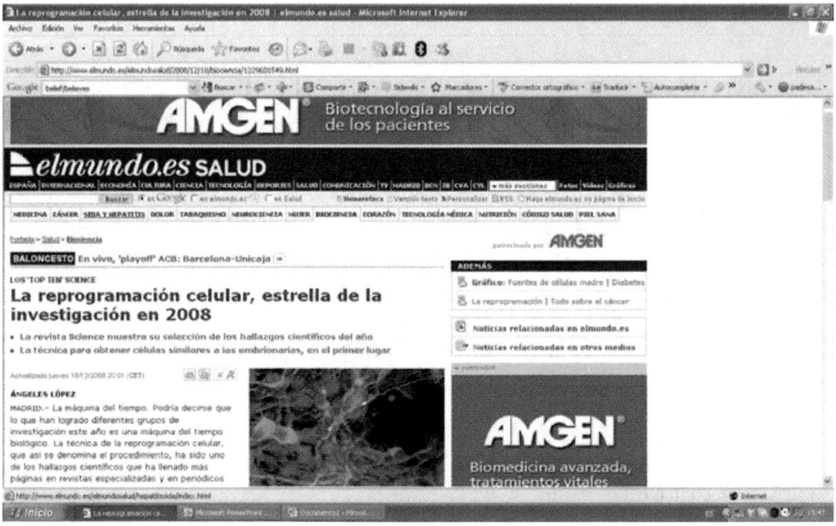

Figure 2. Screenshot of *WebSalud.*

The main formal and compositional features, which are not only textual but include other aspects such as background colours, etc., can be summarised as follows:

The *Salud* SPA has	The *WebSalud* SPA has
▪ Grey background which could give the idea of a static article. ▪ Big red headline ▪ Name of the editor in grey ▪ Small image in grey ▪ Text in verdana 8.5 pt ▪ One hyperlink which takes one to *Science Magazine* homepage	▪ White background ▪ A big blue headline ▪ Grey subheadings ▪ Last uptdated ▪ Name of the editor in bold ▪ Place where the article is written ▪ Image ▪ Text in Verdana 12 pt ▪ Many underlined hyperlinks in blue which direct to other articles from *WebSalud*, supplement *Salud* or the health section of the newspaper ▪ Emphasised information in bold ▪ Icons/buttons to print, send to a friend, zoom in and zoom out, correct something, recommend to someone, receive information in your cell phone, share this article with people in Skype, Twitter, etc.

Table 1. Formal and compositional features of *Salud* and *WebSalud* SPAs.

According to the observable differences between the *Salud* SPA and the *WebSalud* SPA, it can be inferred that the *WebSalud* SPA makes greater use of the potential affordances that are inherent in an online text. In other words, it can be noticed that the *WebSalud* SPA is characterised by a higher degree of dynamism because its content takes advantage to a greater extent of the possibilities offered by the online format. For example, different colours for headings, sub-headings, images, etc. are used in the *WebSalud* SPA as opposed to the *Salud* SPA, where red and grey are the only colours used. Another example is the greater number of hyperlinks given in the *WebSalud* SPA in contrast with only one given in the *Salud* SPA. Furthermore, the links in the *WebSalud* SPA not only lead the reader to another source but also to other related articles published in the same newspaper. Finally, the way hyperlinks are assembled in the *WebSalud* SPA guarantees more coherent and comprehensible information.

Insights into the reasons underlying the greater degree of adaptation of *WebSalud* SPAs to the computer-mediated medium have been gathered thanks to the information collected in interviews with journalists, which are commented upon in the next section. To investigate this aspect, in the course of the interviews special attention is given to the process of planning and writing articles, their new semiotic characteristics, and their effectiveness in the presentation of scientific knowledge.

2.2. The process of planning and writing WebSalud SPA

It is commonly thought that every editor has a virtual reader on their mind when they start writing; however this is not always the case. In the Health Department of *El Mundo*, editors write for both *Salud* and *WebSalud*. Relevant topics are distributed among editors, who are sometimes aware of the specific public they are addressing – the *Salud* readers or the *WebSalud* readers. However, as López (2009)[1] points

---

1 Here 'López (2009)' refers to interviews conducted with Ángeles López, coordinator of the Health Department of the newspaper El Mundo, which have not been published elsewhere.

out, even when a scientific of technical article has been originally assigned to *Salud,* it may eventually be destined to publication on *WebSalud* if good audiovisual support is available. López states that although a current topic such as 'stem cell therapy' is of interest to both *Salud* and *WebSalud,* as stem cells are involved in the treatment of incurable disorders and in the regeneration of body tissues which are of general interest, the topic is better suited to *WebSalud,* because it allows editors to link all the resources and articles which have already been published in the online magazine.

By converse, the printed version is limited in space, and cannot rely on extensive inter- and hyper-textual tools to provide a complete picture of the issue. Thus, articles which offer good opportunities for hyper-mediated cross-referencing are better suited to *WebSalud,* and the decision to publish a piece in the web or print version of the magazine is not necessarily one that is made in advance, but is sometimes prompted by contingent factors which emerge during the writing of the article.

López (2009) insists on the usefulness of hypertextuality in *WebSalud* as a tool for linking all related articles and resources. The impact of hypertextuality[2] on a text is due to the fact that it leads to intertextuality.[3] Many scholars (Askehave/Ellerup Nielsen 2004: 15; Garzone 2007: 23) have highlighted the relative freedom that readers can enjoy in a hypertextual system, which allows them to actively construct their own reading path. "In this environment the conceptualization of one text depends on its relation with other texts" (Garzone 2007: 24). The non-linear reading mode can help readers find information after their own interest and desire, without having to follow a preconceived reading path.

---

2 According to Nelson (1981: 2) hypertext is defined as "a series of text chunks connected by links which offer the reader different pathways". Garzone (2007: 22) states that hypertextuality is indeed the constitutive characteristic of Web text.

3 Derived from the Latin *intertexto,* intertextuality is a term first introduced by French semiotician Julia Kristeva in the late sixties; at that time, the concept was applied to traditional printed texts. Nowadays, it is a common characteristic of Web communication, where the reader is able to shape the text's meaning through hyperlinks.

On the other hand, *WebSalud* is also characterised by very fast publication rhythms; as López (2009) points out, the writing process for this section has to be very rapid. Due to the combination of immediacy and easy access to information, users can constantly access new data and connect it with information they already know. The capacity of online journalism to keep users up-to-date on their chosen topic is greater than that of printed journalism, which publishes news daily or weekly without being able to relate it to previously published materials.

2.3. Potential readers of WebSalud SPAs

López (2009) recognises that the general public does check the Internet for information of topical relevance, but this search for information is not usually accurate, and is frequently conducted in a hurry. She adds that online readers do not entirely overlap with the average *Salud* readers, but rather comprise different categories of users. In fact, the online reader can be 1) a reader interested in a specific article of *WebSalud*, 2) an online reader searching for information about some topic or news or 3) an accidental reader that bumps into the article looking for something else. In order to attract the attention of every category of online reader, editors from the Health Department of *El Mundo* use strong colours in headings, subheadings, images, bold type letters and icons to help the reader read, print or share the information with other people.

2.4. Effectiveness of presentation

López (2009) also says that the online reader's profile should be taken into account in order to achieve effectiveness in presentation. Editors writing for the *WebSalud* focus on finding ways to draw readers' attention to texts. To do so, they use the following means:

1) Highlighting headings and subheadings by means of bold colours. Using coloured text in headlines is appealing because colour is a magnet to the eye and creates a feeling of excite-

ment. Above all, high-intensity colours on a white background are very attractive for readers.

2) Emphasising findings and conclusions by using bold type letters. By these means, editors make sure that the reader always gets a general idea of the content, even when he/she has no time or interest to read the whole article. The affordances that the new medium offers are used by these editors to adapt their writing to the high-speed society we live in.

3) Restoring multimodality to texts. The Internet has brought a resurgence of the visual image into our everyday communication, where text and images are ubiquitous. In fact, "images help viewers understand and picture the information" (Palacios/Diaz 2009: 36), and are therefore particularly effective in science communication.

4) Using innovative language such as metaphors. This device conveys ideas that are otherwise difficult to express. Newmark (1988: 104) points out that the purpose of a metaphor is twofold. On the one hand, its referential purpose is to describe a mental process or state, a concept, an action or an object more comprehensively and concisely than it is possible in literal and physical language. On the other hand, its pragmatic purpose is to appeal to the senses, to interest and to surprise. The first purpose is cognitive and the second one is aesthetic. Newmark considers both important and, indeed, he states that both should be reached when using a metaphor. These purposes are achieved through the mapping of a source cognitive domain (e.g stem cells in example a) below) onto a target cognitive domain (e.g. astronomy) as we can observe in the following examples from the *WebSalud* (a translation is provided to ensure the understanding of the concepts):

a) Las *células madre* se han convertido en una entidad cada vez más familiar para todo el mundo…La revista 'The New England Journal of Medicine' dedica esta semana un gran espacio a estas *'estrellas'* y publica los resultados de tres estudios sobre este tema… (López 2008a)
Cells> stars

b) …si la *célula* fuese un *bombón relleno de guinda* (ADN del núcleo) *y licor* (ADN mitocondrial), este equipo ha logrado sacar la guinda de un ovocito que tenía mutaciones en la mitocondria y transferirla a un nuevo bombón de chocolate con el licor sano, que había sido vaciado previamente de su núcleo. (Valerio 2009)
Cells> filled chocolate with morello cherry (nucleus DNA) and liqueur (mitochondrial DNA)

c) "Las *células madre adultas* son una gigante *caja negra*", setencia Sangiorgi. (López 2008b)
Adult stem cells> black box

d) Desde que hace alrededor de siete años la comunidad científica descubrió el potencial médico de las *células madre embrionarias*, los investigadores no han dejado de buscar fórmulas alternativas para obtener estas *'llaves maestras'*, capaces de convertirse en cualquier tejido u órgano del organismo. (Valerio 2005)
Embryonic stem cells> master keys

5) Using a different design for special issues on topics such as stem cells,[4] Alzheimer,[5] cancer,[6] etc. In this way, editors try to make the reader notice at a glance the relation between articles of a certain section or special topic. Under a general heading, such as 'special on stem cells', a list of related issues is given: 'a short but intense story', 'the secret is on flexibility', 'cell sources', etc. Each of them leads to a small article that summarises the key points.

3. Conclusions

The analysis conducted in this chapter has helped shed some light on current developments in science journalism in Spain. The data considered – albeit limited in scope – make it nonetheless possible to identify a number of features which may be expected to play a

4 See <http://www.elmundo.es/elmundosalud/especiales/2002/11/celulasmadre>.
5 See <http://www.elmundo.es/elmundosalud/especiales/2004/04/alzheimer>.
6 See <http://www.elmundo.es/elmundosalud/especiales/cancer/introduccion.html>.

relevant role in the future evolution of the genre. These features can be summed up as follows:

- Greater importance is being attributed to science in Spanish newspapers nowadays. Indeed, science journalism in Spain is becoming more effective and stronger.
- As online specialised sections are still in the process of becoming established, there are still no dedicated editors for them.
- Online SPAs are being adapted to a fast-changing society. *WebSalud* SPAs are written fast due to the need for immediacy and the profile of the online reader.
- Opportunities to connect articles related to the same topic are increasing. *WebSalud* editors use hypertextuality and a special design to group them.
- There is a need to use a wide variety of stylistic devices in order to attract and retain the reader's attention and interest. Popular scientific editors from *El Mundo* tend to apply a combination of linguistic and semiotic resources (colourful headings, hypertexts, images, bold type, functional icons, etc.) to this purpose. Images give insights into the topic and bring up opportunities for interpretations beyond the explicit content given by journalists.
- Analogy is frequently relied on to explain scientific notions, drawing especially from the reader's background; this is an effective way of re-elaborating complex concepts as it enables readers to grasp scientific notions even while staying on familiar ground.

MIN-HSIU LIAO

A Diachronic Study of the Q&A Column in a Popular Science Magazine

1. Introduction

Popular science writing is relatively young in Taiwan, and the first widely circulated popular science magazine, *Ke Xue Yue Kan* [*Science Monthly*], was launched as recently as in 1970. Since then, genres in popular science writing have undergone a process of constant evolution. One of the most powerful forces driving genre change in this domain has been the interaction between expert scientists and lay readers. In this chapter, our aim is to demonstrate how the relationships between genre users can lead to changes in generic features over time. Since our focus is the interaction between scientists and members of the public, we have selected the Q&A columns as our data source, where such interaction can be seen most clearly.

2. Q&A column as a genre

Bhatia (2004: 23) defined genre as:

> language use in a conventionalized communicative setting in order to give expression to a specific set of communicative goals of a disciplinary or social institution, which give rise to stable structural forms by imposing constrains on the use of lexico-grammatical as well as discoursal resources.

To be able to claim that the Q&A columns represent a genre, it is necessary that they meet the above criteria. At first glance, it may seem fairly easy to say what these criteria are. The genre includes interaction between two parties: questioner and answerer. The communication takes place in magazines, and the columns will be read by all readers. The communicative goal for the questioner is to raise science-related questions, and for the answerer, is to provide solutions to the questioner's problems. The structure is stable in that one question is followed by one answer; i.e. there is only one dialogic exchange going on at any given time.

The column also shows regular textual conventions. Similar to the format of a letter, questions and answers usually begin with a salutation (address terms such as *Editor*, *Mr. Chen*, etc.). In the main body of the text, readers can expect a knowledge deficit to be identified by the questioner, or knowledge to be supplied by the answerer. Optional elements may include compliments to the magazines (from the questioner), a brief account of personal background (from the questioner), or compliments from the answerer on the knowledge-seeking spirit of the readers. At the end, the letter writer signs his/her name.

However, a closer look at the genre reveals a deeper complexity. As in other types of letter writing, the questioner in the column addresses the recipient(s) (usually the editor, or a specific scientist), and the answerers address the questioner. The correspondence may seem private, but is public. The letters may be addressed to a particular recipient, but all the magazine readers can read the correspondence, and the letter writers are clearly aware of the existence of the wider audience.

Previous studies on open letters, such as letters to the editor or letters to advice columns, have all highlighted how the linguistic and discursive practices of such genres are affected by this kind of "multiple dialogues among multiple agents" with "a diversity of goals" (Kreuz/Grasser 1993: 66).

We can analyse the dialogues taking place in the Q&A column on several levels. First, they can be regarded as dialogues between individual questioners and answerers, each with a simple goal: to seek information and to give information. At the same time, the questioner and the answerer may also address the public. A questioner may want to send his letter to a magazine for several reasons, such as to draw the attention of other readers to an important issue, or simply to "get into print" (Kreuz/ Grasser 1993: 66). On the other hand, the answerers may also be addressing the wider audience with a different agenda, such as to demonstrate their expertise or authority (Kreuz/Grasser 1993: 66).

Another agent involved in this multiple dialogue is the editor of the magazine. A hidden agenda behind this activity of letter exchange in magazines is to attract and involve readers, thereby hopefully making the magazine more appealing and thus increasing sales. This agenda may also be referred to as 'the institutional goal' (Locher/ Sebastian 2006: 73) of this column. Therefore, not only are the generic features of this genre shaped by questioners and answerers, but significant textual constraints are imposed by the editors.

In light of this complex picture of multiple genre users and layers of communicative purposes in a popular science magazine, if we want to explore changes in the genre over the decades, we need to adopt a holistic model of genre analysis which will consider factors in a pragmatic context (the interaction between questioners and answerers), in a professional context (the interaction between experts and lay readers, and the intervention of editors), and in a social context (the interaction between scientists and the public in general).

3. A model of interaction between genre users

To account for the entire range of interaction in the Q&A columns – with multiple agents and multiple objectives in only one exchange of written dialogue, our model of investigation is adopted from critical genre analysis as proposed by Bhatia (2004). Genre studies used to place emphasis on describing grammatical features or moves in texts, but more recently, genre researchers have shifted their interest to explore genre as "a powerful, ideologically active, and historically changing shaper of texts, meanings, and social actions" (Bawarshi/ Reiff 2010: 4). In this chapter, we seek not only to describe the Q&A

columns, but also to explain and interpret how they interact with development in popular science, both socially and historically.

Bhatia (2004: 155) described genre analysis as a "multidimensional, multidisciplinary and multi-perspective" task, and this is particularly true when our data also involve such complex multiple interaction. This model divides the investigation procedure into three steps, each focusing on a different space with its own focus:

- textual space: lexico-grammatical features, rhetorical moves, discourse strategies, intertextuality, etc.;
- socio-cognitive (tactical and professional) space: correlation between text-internal and text-external factors, participant relationships and the way they use the genre;
- social space: social identities, social structures and the functioning of social institutions through discursive practices.

The merit of this critical model, which offers a comprehensive and detailed method of investigation and description, is its emphasis on professional and institutional practices, acting as a bridge between text and social context. Bhatia (2008a: 162) calls for genre researchers to include in particular "institutional voices that form part of the wider professional practice." In this study, we will also draw on the role of institutions in shaping the diachronic changes of the Q&A column.

Having divided our structures of analysis into three hierarchical spaces, we ask ourselves what specific research question is to be answered in this study. Hyland (2004), Bhatia (2004) and many other genre researchers offer a long list of research questions that might be helpful in exploring genre practices. In order to illustrate the relationships between genre users (including writers, readers and the other agents involved) in the textual, socio-cognitive, and social spaces, we have decided to focus on one question: how do the relative relationships between genre users contribute to genre change over time?

The relationships between genre users play a significant role in understanding a genre. In the guidelines for genre researchers, an exploration of genre users is always on the suggestion list. Hyland (2004: 195-196) suggests that one of the aims that genre analysis seeks to achieve is to "discover how the genre relates to users' activities". Fairclough (2003: 70) also raises the following two questions: "what are people doing, what are the social relations between them?" Similarly, Bawarshi and Reiff (2010: 4) regard understanding "what reader/writer relationships genres maintain" as a crucial part of genre knowledge.

To illustrate further how changes in one space can lead to and be influenced by changes in other spaces, we conduct a diachronic study of this column over a span of three decades, from 1970 to 1999, which marks the most significant development of this genre in Taiwan, from its introductory stage to its maturity.

In the textual space, we place emphasis not only on who the questioners and answerers are, but also on their relative relationship and status; in other words, how they see each other, and how this mutual assumption influences their use of language. The pattern of language in relation to relative status of genre users is our focus at the textual level. The specific lexico-grammatical features we investigate in this space are terms of address.

In the professional space, we investigate the roles of questioners as lay people, and answerers as experts, and highlight interventions from editors in particular. As mentioned above, the 'open letters' correspondence in the columns may seem private and individual, but the procedure – from the moment when the letter was received, and answered, through to publication – actually involves a series of editorial decisions. The interaction between these two roles (writers and editors as experts and readers as lay people) and the institutional constraints are our focus at this level.

Finally, we expand our discussion to the interaction between scientists and the public in the social space. Interaction between questioners as lay people and answerers as experts is not stable (as we will discuss in our diachronic analysis below), because the way they interact is largely dictated by social practice. How science is valued in society influences how scientists as a profession are perceived by the public, and this contributes to the interaction between the group of questioners and answerers in a science magazine.

The theoretical framework to be used for our analysis is summarised in Table 1 below.

Space	*Interactants*	*Data to be investigated*
Textual	Questioners and answerers	Address terms in letters
Professional	Experts, lay people, and editors	Editors' words, announcements of call for letters, comments on the columns in the magazine,
Social	Scientists and the public	Media reviews on development of science and the role of scientists

Table 1. The theoretical framework used to investigate Q&A columns in magazines.

4. The corpora

For the purpose of this study, three corpora were compiled, including all the correspondence in the Q&A column in *Science Monthly* from 1970 to 1999. The three corpora are presented in Table 2 below.

Corpora	*1970s*	*1980s*	*1990s*
Word count	104,230	6,146	52,064
No. of questions	176	52	142
No. of answers	212	53	99

Table 2. The corpora of Q&A column in *Science Monthly*.

The Q&A column across the three decades maintained a similar format until the late 1980s, then a significant change occurred in the 1990s, which will be our focus of discussion in the three spaces. In the 1970s and the 1980s corpora, the answers to readers' questions were provided by the magazine – the editors, or scientists who had been invited to contribute by the magazine. However, in the 1990s, the magazine readers were also given the opportunity to answer questions, and were rewarded with prizes if their answers were published. This move had a significant impact on our research question, which focuses on how interactants perceive each other, and we will discuss this change more in depth below.

5. Interaction between questioners and answerers

Participants in Q&A columns, as interactants in other forms of communication, have to make assumptions about their addressees and their relative status. Linguistic choices in their interaction can tell us a significant amount about their projection of relationship, and among all the linguistic resources, "the choice of terms of address is one of the first of these recognized by sociolinguists" (Scollon/Scollon 1995: 35).

Table 3 shows the occurrences and percentages of different salutations used by questioners at the beginning of the letter to greet their addressees.

Corpora	*Address terms*	*No.*	*%*
1970s	Mr. Editor [編輯(者)先生]	105	56
	To whom it may concern [敬啟者]	36	19
	(Readers' columns of) Science Monthly [科月(讀者信箱)]	21	11
	No salutation	13	7
	Specific Editor	11	6
1980s	Mr. Editor [編輯(者)先生]	36	69
	To whom it may concern [敬啟者]	7	14
	Specific Editor	4	8
	No salutation	2	4
	Editorial section [編輯部]	2	4
	(Readers' columns of) Science Monthly [科月(讀者信箱)]	1	2
1990s	Editor/ chief editor/ old editor [編輯/主編/老編]	61	43
	Everyone (and other variations such as: all/ old brothers and old sisters/every wise man) [大家/各位/大哥大姐/各位智者]	25	18
	(Readers' columns of) Science Monthly [科月(讀者信箱)]	18	13
	No salutation	14	10
	Editorial section [編輯部]	12	8
	Editor and readers [編輯和讀者]	6	4
	Readers [讀者]	2	1
	Professor [教授]	1	0.7

Table 3. Salutations in questioners' letters.

It is interesting to consider to whom the writers should address their letters. As we have already pointed out, these letters are also read by the editors and all the readers of the magazine, so the choice of terms of address is not simply a lexical choice but an interpersonal choice – to whom questioners want to talk and how.

In the 1970s, the address form most commonly used by letter writers is *Mr. Editor*. Addressing people by their professional title is a common practice in Chinese; doing so acknowledges the addressee's social role and shows respect to him/her. The second most frequent address form is *(Readers' column in) Science Monthly*. In this case, it is the column or the magazine being addressed rather than the editor. This choice shows a detached relationship, and implies a great distance between the two parties. Finally, *to whom it may concern* is also used a few times. This term of address indicates a great social distance and is used in formal occasions, usually when the individual being addressed is unknown.

In the 1980s salutations follow a pattern similar to those used in the 1970s. Most letter writers chose to address the editors, or used the detached form *to whom it may concern*. In this period, it seems that addressing a person is preferred to addressing an institution, as we can see that the use of terms of address like *science monthly* or *editorial section* declines substantially in the 1980s.

In the 1990s, the question of whom the letter writer should address his/her correspondence to became more complicated, because, as we mentioned in the data description, the answers can come from all readers who can propose solutions to the questions. When a letter is written, the questioner does not know to whom the letter is addressed, but it is clear that editors and magazine readers will certainly read this letter. Therefore, we see a variety of choices in this period. Most letters are still addressed to editors, but the percentage is not as high as in the 1970s and 1980s. In the 1990s, 18% of the letters address the general public such as *everyone* or *all*. A similar percentage of letters is addressed to the editorial departments, the magazines, or has no salutation. A small percentage of letters are addressed to both the editors and the readers, and some are even addressed to the readers only. One letter uses the address form *professor*, presumably as a general term for science experts.

The frequencies of these address categories aside, it is worth making reference to the formality of the address terms. In the 1970s and 1980s, the address forms follow a very conventional pattern, usually *Mr. Editor*, or *Science Monthly*, but in the 1990s, we see a range of informal and unconventional choices of address terms, such as kinship terms like *Old Brothers and Sisters in Science Monthly*; creative ones like *every wise man* or *seniors with insightful answers*; and *old editor*, an informal way of addressing the editor. In many cases, the address forms are as simple as *hi everyone*, or *dear readers*. The assumption behind using these non-conventional terms is that there is no need to use formal and detached forms of address for the people who are going to respond to questions. The answers could be delivered by anyone, not only one particular privileged group. Scientific knowledge does not come from a particular group only, and the people who have the knowledge to provide the answers do not necessarily have to be addressed differently from other people. The power relationship between questioners and answerers is presented as equal in these cases.

Let us now consider the choices of address terms made by the answerers. Table 4 shows the occurrences and frequencies of salutations in the replies in the 1970s and 1980s. The answers in the 1990s corpus do not contain salutations, because these answers are presented in the magazine with an introductory line from the publisher saying that *Reader [name] would like to respond to questions from reader [name]*. We will discuss below how having no address term in the answers also projects a kind of relationship between interactants in the 1990s corpus.

Corpora	*Address terms*	*No.*	*%*
1970s	Mr./Ms. [先生/小姐]	112	53
	Student [同学]	46	21
	No salutation	35	17
	Reader [讀者/友]	19	9
1980s	Mr./Ms. [先生/小姐]	25	49
	Reader [讀者/友]	11	22
	Student [同学]	9	18
	No salutation	6	12
1990s	N/A		

Table 4. Salutations in answers.

From Table 4 we can see that in the 1970s and 1980s, the most common form of salutation for addressing the readers was *Ms./Mr.* with surname, first name, or full name. As with the choices of salutations in the questions, we can see that the relationship between answerers and questioners during these two decades is marked by distance and formality.

However, Table 4 also shows a very interesting trend: the use of *student* to address the readers. If the questioners are identified as students, either through their self-reference or through the content in the letters, the answerers often address them as *student* with their surnames, first names, or full names. By addressing the readers as students, answerers perceive their relationship with the questioners as a teacher-student interaction. The address term *student* is a linguistic feature typical in an academic interaction, and implies the hierarchy of a teacher in a higher position, having authority.

Overall, in the 1970s and 1980s, linguistic choices are characterised by the typical generic features of interaction in an academic context – a formal relationship between academic colleagues or between teachers and students. We will see more evidence of the 'colonization' [1] by academic genres in this period later in the discussion.

Let us now look at the 1990s corpus. The change in the forum in this decade influences the use of address terms in letters and also self-reference. For one thing, as we have mentioned previously, the answers are presented not as private letters but rather as pieces of information introduced by the editors' words, so there is no direct greeting (salutation) by the answerers. Therefore, unlike in the 1970s and 1980s corpora, the answerers in the 1990s do not have an opportunity to show us how they project their relationship with the questioners. We may say that in this decade, the mutual distance between questioners and answerers is largely decided by questioners, through the salutations they use to address the answerers. In fact,

1 Genre colonisation is defined in Bhatia (2004:58) as a process of invasion of one genre by another, "often leading to a creation of a hybrid form, which eventually shares some of its genre characteristics with the one that influenced it in the first place."

many of the people who answered questions in the 1990s were the same people as those in the previous two decades – university lecturers or science experts – but they had lost one of the most important linguistic mechanisms, i.e. the possibility of choosing which salutation to use, in order to manage their interaction with questioners.

We would argue that by losing the choice of terms of address, these answerers have lost one of the most effective linguistic tools to define the position of their addressee in the interaction. We do not mean that all the answerers want to posit themselves in a higher position than that of the questioners – since in the 1970s and 1980s a variety of different forms were used. However, it is clear that the interpersonal relationship projected by the choice of address terms is filtered out in the editorial process, and the informative function of the letter is foregrounded by the editor's introductory line. The publisher's intervention in the format of the column has a significant impact on the interactive features in the questions and answers, which is what we are going to discuss in the next section: the professional space.

6. Interaction between experts, lay readers and editors

In the previous section, we investigated how the use of address terms reveals interaction at the level of individuals. However, Graham (2001: 2) argues that the areas of focus identified in critical genre studies, such as generic instability and genre hybridity, enable us to investigate changes particularly "at the level of institutions". Therefore, we now further pursue our research question by looking from a different angle at what kind of relationship the institution expected to be constructed between questioners and answerers in each decade. The focus here is the interaction among three groups: experts, lay readers, and editors.

Announcements that call for letters and promote this column give us an insight into the editorial views. These announcements encouraged the writing of letters and had a potential impact on the interaction in the columns.

In the 1970s, the Q&A column in the magazine was usually accompanied by a statement (shown below) to explain what they expected from readers.

> We particularly welcome readers who raise questions which they have thought over. In your letter, it is better to write down your reasoning process and the difficulties encountered, as a reference to answers and joint investigation.
> The answers provided in this column exclude exercises from textbooks and those that can be easily found in high school textbooks – unless that question is of special value or involves difficulty beyond the scope of textbook.
> To those questions where reference can be easily found, we provide only brief answers.[2]

This announcement represents the 'institutional voice' – the person(s) with whom the letter writers assume they are interacting. The first personal pronoun *we* is a collective term for all the experts and editors who stand behind this column and give answers. Therefore, what kind of 'institution' does this voice represent?

The authoritative directive tone and lexical features deployed in sentences such as "write down your reasoning process and difficulties encountered", "exercises from textbooks", and "questions from which references can be easily found", are typical of an academic institution. These words sound familiar as examples of what a teacher would say to students – to encourage them to think, and to say that s/he is not going to answer simple questions. Although the text states that they want to exclude textbook-related questions, by positioning itself as something additional to a school setting, this column can be seen as an extension of science education in schools.

Considering the professional identity of the answerers and editors – mostly university researchers or lecturers – it is not difficult to see why the institutional voice of this magazine in the 1970s was influenced by the voice of academic authority, telling readers what kind of questions are expected and particularly what questions are not encouraged.

The influence of editors' authoritative directive tone on readers' letters is obvious. Many readers in the 1970s accommodated their

2 All the quotes from the magazines are our translations from Chinese.

letters to the guidance of the editors, as illustrated by the example below:

> Readers' columns of Science Monthly,
>
> I took my spare time in the Winter vacation to read this science magazine in great details, and I was very interested in the story of *Hanxin Dianbing* in the first issue. After reading it, I was wondering whether the theory of *Sunzi* could be modified in the following way:
> […]
> This modification is only my personal thoughts, without reference to any books. Since I am only a high school student with limited understanding of scientific knowledge, my reasoning inevitably contains errors or imperfection. I sincerely request the editor to provide me with instructions.
>
> Best wishes,
> Reader Lin Zhong-Yuan

This letter demonstrates how the reader responded to the editors' suggestions by showing that the question came from the first issue of this magazine and not from topics covered in high school textbooks, and also showing evidence of independent thinking by proposing a modification to an existing theory, as they were encouraged to do. The last paragraph in the letter also shows how this reader perceived his relationship with the editors – a student with limited knowledge and an expert who could provide him with instructions.

We can say that questioners' and answerers' use of address terms such as *teacher* and *student* does not derive entirely from individual choice, but are subject to the constraints of a professional space manipulated by the institution. Although compared with other genres with more rigorous generic conventions, such as legal genres, the conventions discussed here are not binding, in the textual space certain expectations are imposed on and reflected in the linguistic choices made by genre users.

In the 1980s, the numbers of letters from readers dropped and this column ceased to appear on a regular basis. Here below is a different announcement made by the editor to call for letters.

> Readers' columns […] provide a channel of communication between Science Monthly and readers […]. We sincerely hope everyone can make use of this channel, but texts should be kept short and unfocused conversation should be avoided.

Compared with the announcement in the previous decade, this text has fewer linguistic features related to the school context, but still retains an authoritative tenor, telling readers what the editorial staff expect from the readers (to make use of this channel) and what they do not want to see (unfocused conversation).

Towards the end of 1990s, a new format of the Q&A column was launched, from which we can see a significant change in the institutional voice:

> Science Prize is an area for readers to pose questions and exchange knowledge. Please dig out interesting questions from your daily life or study. Exchange thoughts through discussion. Do not be afraid lest your question is too stupid, and do not be afraid lest your answer is not professional. We sincerely hope that readers will react actively and participate in cultivating the seeds of science.

If in the 1970s institutional discourse can be seen as being colonized by academic discourse, this text, in some way, exhibits some features that are similar to those of promotional genres. The purpose of this column, as stated in the magazine, is to provide a platform for the communication of science knowledge. This institutional goal has remained the same for three decades. However, we cannot ignore a higher goal for a commercial publisher – commercial profits. We mentioned that in the 1980s, the number of letters dropped, which was largely related to a dip in sales. Therefore, in a wider sense, the new format of the Q&A column launched in the 1990s can be seen as serving the commercial purpose of attracting readers and increasing sales. The new title of this column, Science Prize, is typical of advertising discourse – if you respond to this advertisement, you will have the opportunity to win a prize. Not only is this sensational in linguistic terms, but in practice the publisher did provide prizes to magazine readers who participated in this column. Also, the text shows a strong intention to build solidarity with readers by encouraging them not to be deterred by questions which may seem stupid, and by inviting them to contribute to the development of science. The institutional voice is not as authoritative as in the 1970s, but has become persuasive, just as in promotional genres.

Influence of this trend of commercialization can be seen from letters sent by readers in this decade:

> Great Editor,
>
> I heard from people that dogs are colour blinded. They do not recognize others by colours but by smell. Is this true? […]
>
> Little Cindy

This letter demonstrates different interpersonal features from those shown in the example from the 1970s. The reader did not attempt to persuade the answerer by providing evidence of their independent thinking or denigrating their position and therefore praising the editors'. Also, the forms of address and self-reference were more informal, as observed at the textual level.

The change of address terms in late 1990s, therefore, seems to be a shift led by the publisher to impose a different convention on the genre users. However, if we go beyond the institutional level, the truth may be that the decision made by the institutions is passive, and is actually a response to what lay readers want. In the next section, we will move our focus to the social space.

7. Interaction between scientists and the public

The shift of institutional voice from authoritative to persuasive is closely tied to changes in the relative positions of professional scientists and the public. The year 2010 marks the 40th anniversary of *Science Monthly*. In a speech reviewing the history of this magazine, one of the editors observes that now taxpayers have realized that scientists are merely one type of profession in society, and they do not have any privileged status to enlighten the public (Wang 2009).

When *Science Monthly* was first published in 1970, Taiwan was closely tied to an ideology of scientific authoritativeness (Jiang 1990). This magazine was launched not only for the purpose of spreading information about science but, more profoundly, it marked the ideals and passion of a group of overseas Taiwanese students who wanted to strengthen the country by educating the public with advanced science knowledge from the West. One of the founders described themselves as "a group of students who are away from home, love their country, and are willing to contribute their knowledge to the country" (Lin 2009). Huang (2010) criticises these scientists for presenting themselves as a cultural elite, intending to educate the public who do not have enough science knowledge. The deep belief that introducing science to the public is the only way to develop the country is reflected in its authoritative institutional voice and in the colonization of academic genres by teacher-student relations in the 1970s and 1980s.

The public in those days also greatly admired professional scientists, particularly those who received an education in the U.S. Not many people in Taiwan could afford to study abroad at that time, so overseas students were highly privileged. The editor's opening address in the first issue clearly reflects how the public perceived overseas Taiwanese scientists:

> Because of the portrait of a few Taiwanese students studying in the U.S in newspapers or magazines, many people have an overwhelming feeling of admiration for 'overseas students in the U.S.', as if they were a treasure that they dare not touch, lest it should fall on the floor and break. Therefore, a magazine launched by such admirable 'overseas students in the U.S.' will surely be outstanding.

Although the editor went on to deny this statement, and emphasized the team's inexperience in popular science publications, they had in fact presented themselves as a group of humble people who just wanted to make a contribution to the country. However, this opening address reflects how this group of scientists was perceived in Taiwanese society at that time.

In the late twentieth century, as in other parts of the world, people were gradually witnessing the side effects of scientific development. The power of science and the status of scientists were challenged. The anti-nuclear movement in Taiwan in the 1980s further increased people's doubts about whether there is one scientific truth. Against this distrust, scientists could not simply tell the public what to do, but they had to persuade the public to trust them. In the Q&A column the scientists' authority is demystified through the mechanism that allows everyone to give an answer to questions, and the terms of address become less formal and detached. The myth of 'one scientific truth' is also broken by the design of the column, which makes it clear that one question can lead to several answers and each represents a different view. These changes in textual space and professional space not only reflect social changes, but at the same time, they consolidate changes in the social space.

8. Conclusion

This chapter has addressed one specific question in genre studies: how do the relationships between genre users contribute to genre change over time? Drawing from methods proposed by Bathia (2004) in his critical genre analysis, we found answers with regard to three spaces: textual, professional, and social. At the textual level, the data shows that across the three decades, address terms and self-references used by questioners and answerers moved from more formal to less formal choices. Moreover, in the 1990s, questioners began to address other readers as well, moving from lay-expert exchanges to multi-agent dialogues. Changes at the textual level are related to changes at the institutional level. In the 1970s and 1980s, editors guided readers as to what to ask and how to ask, as if they were students and teachers; but in the 1990s, editors had to encourage readers to send in letters, and even to motivate them with rewards. These changes were set against changing attitudes towards science and scientists in the Taiwanese society across the three decades, from admiration to prevalent distrust.

This study has shown that changes in these three levels are not necessarily influenced from the top down or bottom up, but each layer of practice may influence and at the same time be influenced by other layers, and gradually, through the interaction between different groups of genre users, give rise to recognized and agreed genre conventions in the discipline.

Institutional and Business Discourse

ANNARITA TAVANI

FAO's Food Insecurity Reports: An Evolving Genre

1. Introduction

This study rests on the recognition of the increasingly important role played by various types of institutional communication in politics and society, a fact that has recently drawn the attention of linguists. Institutional communication is a vast area of study, considering the variety of texts it involves and the diverse contexts of production and reception which characterize it. Within this area, an interesting subgenre whose popularity is on the increase is represented by annual reports drafted by international organizations. In particular this chapter examines the ten FAO reports on the *State of Food Insecurity* (*SOFI*),[1] issued yearly by the Food and Agriculture Organization of the United Nations from 1999 to 2009, with a view to identifying possible indications of evolution in the genre over the decade since the first issue was published. By virtue of their apparently high degree of stability in discursive structure, coupled with what seem to be significant changes in discursive practices, the *State of Food Insecurity* reports may be taken as an example of institutional discourse which is worth studying both synchronically and diachronically within an ampler research project on institutional communication.

In 1996 leaders from 186 countries met at the World Food Summit in Rome and made a solemn pledge: to halve the number of hungry people in the world by the year 2015. Three years later, in 1999, the first report on the *State of Food Insecurity* was issued in English, French and Spanish on the FAO website and published in

1 Available at FAO <http://www.fao.org/publications>.

print by the Food and Agriculture Organization of the United Nations. In 2000 world leaders agreed on eight Millennium Development Goals; the first was to halve the prevalence of hunger (percentage of hungry population) by the year 2015. Since then, an annual report has been issued by FAO yearly, registering progress and setbacks on the way to both the World Food Summit and the First Millennium Development goals. The only exception was the year 2007, when no report was issued. *SOFI* is not FAO's only thematic publication. The other flagship publications are: the *SOFA* report (*State of Food and Agriculture*), the *SOFIA* report (*State of Fisheries and Aquaculture*), the *SOFO* report (*State of the World's Forests*) and the *SOCO* report (*State of Agricultural Commodity Markets*).

The first aim of this chapter is to analyze the distinctive traits of the *State of Food Insecurity* reports and thus define the sub-genre they represent. A second aim is to explore whether signs of evolution can be detected in the discursive practices deployed in the ten issues of the report. Indeed, genre theory conceives genres as dynamic models, since the typified conventions underlying genres are often subject to evolution due to many factors (Berkenkotter 2008a). Therefore a diachronic perspective on genre analysis is of great relevance both for an in-depth understanding of the texts considered and for a clearer definition of the genre they belong to.

In order to discuss the issue of genre definition and genre evolution, the research adopts a double methodological framework, integrating a text-analytical approach (Werlich 1983) with the study of the encoding/decoding context (Bhatia 1993, 2004; Swales 1990), thus aiming to contribute to the generic analysis of the *SOFI* reports. The contextual coordinates of the genre are examined in Section 2, which delineates the communicative situation in which the encoding/decoding process takes place. In this respect, the first point to be discussed regards the identity of the participants, i.e. drafters and recipients, while a second crucial point is the identification of the purpose or purposes of the texts, which help define the genre they belong to. Section 3 analyses the textual structure of the reports, introducing a profiling of the documents in terms of text types – description, narration, instruction – and proceeding to detect signs of change in this pattern over time. The linguistic category of person (Gaynesford 2006) comes under careful scrutiny, with the analysis focusing on a representative sample of text sections, namely the *Forewords* to the ten editions. The choice of the *Forewords* for closer scrutiny is due to the fact that they are like nutshells enclosing the main textual characteristics of the entire reports, including the alternation of linguistic devices typical of different text types. The concluding section of the chapter sums up the findings of the research and envisages prospects of further study on genre-mixing and embedding in the *SOFI* reports.

2. Context and genre

The *State of Food Insecurity* report is the product of FAO's commitment to the highly humanitarian cause for the eradication of hunger around the world. This document, drawn by a team of FAO experts ever since 1999 with the only exception of 2007, has benefited from information and expertise pooled by a number of organizations working together to combat undernourishment in both developing and industrialized countries. The Food Insecurity and Vulnerability Information and Mapping System (FIVIMS), which is the main source of information and statistical data for the various editions of *SOFI*, relies on precious and qualified research work of regional, national and international organizations, operating in the area of undernourishment and malnutrition. Authorship of the report has always been choral in nature and its drafting has always been the fruit of collaborative team work both inside and outside FAO, consisting mainly in the gathering and processing of information. Yet over the years this collaborative effort has increased and authors' identity has turned into a truly global voice. The latest edition of the *SOFI* report considered here, issued in 2009, is the product of the joint work of the Food and Agricultural Organization (FAO) and the World Food Programme (WFP) team. The close to the *Foreword* of this issue of the report clearly states this double authorship, being jointly signed by Jacques Diouf, FAO Director-General, and Josette Sheeran, WFP Executive Director.

This progressive choralization of authorship is matched by the plural character of the reading public addressed in these reports. As a matter of fact, the second edition already reveals the double identity of the audience: the document is targeted at a community of practice made up of experts, namely philanthropists and politicians, while at the same time addressing the general public consisting ideally in every citizen of the world. We shall see in the analysis how this multiple readership affects the design of the reports. It is useful to remember that starting from the first issue, all *SOFI* reports have been published on FAO's website as well as in print by FAO's publishing authority in Rome. It is also interesting to note that the first issue reserved full copyrights, thus prohibiting any unauthorized circulation of the material in the report, whereas the 2000 and all subsequent issues have allowed free circulation of the documents for any purposes other than commercial ones without prior application for permission to FAO's publishing authority. Another interesting fact is the gradual increase in the number of languages in which the reports have been published. This circumstance mirrors the increase in the number of official languages used by FAO. The 1999 report was published in English, French and Spanish, the 2000 edition in English, French, Spanish and Arabic, in 2001 Chinese was added and in the 2008 edition there was the further addition of Russian. The reading public aimed at in the *State of Food Insecurity* report has become ever more global.

A further contextual feature which is fundamental for assigning the *SOFI* reports to a specific discourse genre is their purpose or purposes (Bhatia 1993, 2004; Swales 1990). The presence of multiple purposes, not all of them amenable to single interpretation, indicates genre-mixing and genre-embedding (Bhatia 2004). The first communicative purpose underlying FAO's publication of the *State of Food Insecurity* is the circulation of statistically updated information about the number, prevalence and location of hungry people as well as the depth of their hunger. Intertwined with this purpose there is a further crucial aim at the heart of this report: calling the international community to arms in the battle against undernutritrion and starvation. The description of present reality and the narration of regional and national stories of success or setbacks on the way to a hunger-free world answers the need for dissemination of essential knowledge, while the definition of viable strategies to eradicate hunger aims at spurring the international community to take common action on this global issue. The title of a section in the 2003 edition clearly illustrates this double purpose:

(1) Mapping poverty and hunger to help wipe them off the map (*SOFI* 2003)

Actually, the *State of Food Insecurity* report can be seen as the artful combination of two genres. On the one hand, it shares many of the features which are typical of financial reporting, in particular of the corporate annual report, providing information about results obtained over the past year in the struggle for hunger eradication. On the other hand, the second communicative purpose of the report is very close to that of philanthropic fundraising discourse (Bhatia 2004). Actually, FAO is requiring politicians to both increase development aid and take decisions which may lead to the attainment of the World Food Summit and of the First Millennium Development goals (a 50% reduction in the number of undernourished around the world and a 50% reduction in the prevalence of hunger respectively). Fundraising letters and *SOFI* reports share the same structure: establishing credentials – cf. the section 'About this report'–, enclosing documents to demonstrate the need for action, and finally soliciting response on the part of the addressees, which turns into a call to arms in the battle against undernourishment and malnutrition. This pattern of genre-mixing and embedding is fairly stable throughout the whole series of reports, and provides an anchorage point for their definition as a discourse genre.

3. Text analysis

3.1. Text type profile

The *State of Food Insecurity* reports show a dynamic text type profile, which is the most striking evidence of their complex structure as an institutional genre. The standard report encases several text types ranging from description to narration to instruction, thus modulating the communicative purpose of the various sections. Typically, this report opens with a *Foreword* stating the overall purpose of the document and anticipating its textual profile. There follows a description of the impact of food insecurity on the life of millions of people (*Undernourishment around the World*) and the presentation of a few case studies narrating success and setbacks on the way to food security (*Towards the Summit Commitments*). The last chapter or section (*The Way Ahead*) instructively delineates action which must be taken if hunger is to be eradicated. The titles to the individual chapters are subject to variation and so is the distribution of text types.

These variations reflect the dynamism of the text type profile in the various editions of the report. From a simple description-narration-instruction pattern in the earlier issues, in time there is a movement towards a more complex alternation of text types, often mixing description and narration and interspersing elements of instruction within descriptive or narrative sections. The first report (1999) descriptively introduces the methods for counting and locating hungry people, then goes on to narrate progress and setbacks in developing countries. After a descriptive chapter on the nutritional status and vulnerability of millions of people, the report steers towards its instructive conclusion, featuring strategies of change. The 2000 report shows a similar alternation of text types in that it falls into two clearly descriptive chapters, followed by a narrative one telling the readers about the changes which have taken place in a few sample countries, while the close remains an instructive chapter on work which must still be done. It is interesting to note that the second descriptive chapter begins with an expository section entitled 'The spectrum of malnutrition', which analytically examines the various conditions of malnutrition, ranging from macronutrient to micronutrient deficiency to overnutrition. From 2001 to 2005 *SOFI* reports tend to vary the pattern of text type distribution and reveal frequent overlaps between description, narration and instruction. In particular, the 2003, 2004 and 2005 issues mix description and narration in the first chapters, while the last chapter combines elements of description, narration and instruction. The 2006, 2008 and 2009 issues reveal a prevalence of narration and description in the first and second chapters, while the third and last instructively focuses on prospective change. There seems therefore to be a return to the straightforward structure of information-giving units followed by action-requiring units of the first issues.

The shift from a text type to another is often signalled by a change in tense – from the present simple of description to the past simple of narration to the future forms of instruction – and in person – from the third person singular and plural of description and narration to the inclusive first person plural of instruction. The frequent mixing of descriptive and narrative text types in the *SOFI* reports is due to some fundamental features the two text types share, as they both focalize on factual phenomena – occasionally in the case of narration on conceptual phenomena as well (Werlich 1983). The difference only lies in the context – spatial for description, temporal for narration – which determines the different use of tenses. Instructive texts stand apart in that they focus on future behaviour in both the encoder and the decoder (Werlich 1983). Still in Werlich's terms, in descriptive texts the 'thematic text base' – which offers the basis for 'thematic expansion' – is a simple phenomenon-registering sentence, while in narrative texts the same function is performed by a simple action-recording sentence and in instructive texts by an action-demanding sentence. The partial similarity between descriptive and narrative text types is the reason for their frequent mixing in many chapters and even sections of the reports. Instructive texts tend to be more monolithic, although text type mixing and embedding does occasionally involve instruction too.

A similar artful orchestration of text types is also detectable in the *Forewords*, which epitomize both the textual structure and the participation framework of the entire *SOFI* reports. This is why they have been chosen as the object of a diachronic analysis deploying the category of person, which may help to highlight the distribution of roles between authors and addressees.

3.2. Person in the Forewords

A study of the use of personal pronouns in these microtexts may shed precious light on the participation framework featured in the *State of Food Insecurity* reports. All the *Forewords* see an alternation between the use of the inclusive 'we' – referring to FAO and other international organizations as well as the reading public at large – and the personal 'they' – referring to the population of either the developing or the developed countries. There is also the use of the non-personal third person – both singular and plural – whereas the second-person pronoun is totally absent. This is a sign of lack of reader engagement, which is rather unusual in an essentially directive text. It can be hypothesized that the reason for this is that the directive is in fact self-directed – with the use of the inclusive 'we'. Jacques Diouf, Director-General of FAO, is actually committing himself and the organization he represents to action for hunger eradication. However, by choosing the inclusive first person plural form, he spurs the readers to join common action for a vital humanitarian cause. An example of the role of the inclusive 'we' and of third-person pronouns may be quoted from the *Foreword* to the 2004 edition:

(2) We MUST do better. According to FAO's latest estimates the number of hungry people in the developing world has declined by only 9 million since the WFS baseline period, despite commitments made. More alarming still, the number has actually increased over the most recent five years for which numbers are available. In three of the four developing regions, more people were undernourished in 2000-2002 than had been the case in 1995-1997. Only Latin America and the Caribbean registered a modest reduction in the number of hungry people. We CAN do better. More than 30 countries, representing nearly half the population of the developing world, have provided both proof that rapid progress is possible and lessons in how that progress can be achieved. (*SOFI* 2004)

Against what seems to be a fairly high degree of stability in the handling of person, a few signs of change over the various issues are traceable. It is interesting to note how the first-person singular pronoun 'I', clearly referring to the author of the *Foreword* – namely FAO Director-General Jacques Diouf – only occurs twice, in the sixth and then in the last paragraph of the *Foreword* to the 2000 issue, conferring a strong sense of personal commitment to the Director's words:

(3) I am disturbed to report that we find no significant change for the latest period, 1996-98, compared with the 1995-97 period reported last year. [...] I am convinced we will see the day when FAO ceases to publish a report titled *The*

> *state of food insecurity in the world* because the world will have lived up to its promise to end hunger. (*SOFI* 2000)

The first-person singular pronoun appears in no subsequent issue, where a plural form is regularly opted for. The pervasive occurrence of the inclusive 'we' in the whole series of *Forewords* obviously has the effect of creating a sense of common responsibility and collaboration between author and reading public, both ideally belonging to the world's community of philanthropists and therefore sharing the same principles and objectives. Though undoubtedly pervasive, the number of occurrences of this pronoun varies from issue to issue, revealing a tendency to decline in the post-Millennium editions. A good example of this trend is the *Foreword* to the 2001 report, where the dominant person is the third person, both personal and non-personal, either singular or plural. Here the inclusive 'we' only steps in in the final paragraph to emphasize the possibility and the need for common action against hunger. By contrast, the 2002 and 2004 editions are extremely rich in occurrences of the 'we' form. The two *Forewords* almost entirely consist in a personal appeal to the international community to work together against the common enemy: hunger and its crippling effects on mankind. The 2004 *Foreword* even incorporates the inclusive 'we' in the titles to three central paragraphs: "We MUST do better", "We CAN do better" and "We cannot afford not to do better". From the 2005 edition onwards the use of the first-person plural pronoun in the *Foreword* is on the decline, although it is only in the last two editions that it appears to be almost always replaced by third-person pronouns, with the exception of the close to the *Foreword* to the 2009 issue, where 'we' and 'our' refer to Jacques Diouf and Josette Sheeran and the organizations they represent.

(4) This year's *State of Food Insecurity in the World* is a true collaborative effort between our two organizations, combining our different strengths to create new insights and a publication that benefited tremendously from our joint cooperation. Collaboration with the United States Department of Agriculture on certain parts of the report has also been instrumental and is highly valued; we thank them for their efforts and willingness to share their expertise. (*SOFI* 2009)

The choice of both personal and non-personal third person pronouns, which dominates *SOFI* reports 2008 and 2009, is in accordance with another feature of discourse in later issues of the report, namely technicalization, which will be discussed in the next sub-section. Indeed, discourse tends to be more and more precise and impersonal as it moves from a passionate appeal for concerted action against undernourishment to an accurate study of the problem, which rests on up-to-date statistical data and their correct interpretation.

3.3. Technicalization and popularization of discourse

A further level of linguistic analysis which may be applied to both the *Forewords* and the entire reports is that of the variety or varieties of language chosen by the authors for effective communication. The first issue of 1999 reveals a preference for general English, except for a few mildly technical terms such as 'undernourishment', 'undernutrition', 'food insecurity', 'vulnerability' and 'anthropometry', whose meaning is always clarified by the context. On reading the report issued in 2009, we are struck by a notably different use of language, which is far richer in technicalities, above all from the domain of economic English, and shows a clear tendency towards the de-personalization of discourse, for example through frequent recourse to the use of the passive voice (Gotti 2005). An example may be quoted from the *Foreword* to this edition:

(5) To ensure that hunger is conquered in the years to come, developing countries must be assisted with the development, economic and policy tools required to boost their agriculture sectors in terms of both productivity and resilience in the face of crises. […] Investments in food and agricultural science and technology need to be stepped up. […] Whenever possible, efforts should be integrated and produce a multiplier effect. (*SOFI* 2009)

This massive use of the passive is instrumental to a de-focusing of the reader's attention from the agent and is typical of specialized discourse. Another syntactic strategy which contributes to both de-personalization and specialization of discourse is nominalization, a tendency to convert verbs and sometimes adjectives into nouns, with a consequent semantic unloading of the verbs – often reduced to mere copulas – and a more effective structuring of the sentence in terms of thematic base and thematic expansion (Gotti 2003). The following quotations show the use of this technique, which finally dominates the 2009 issue:

(6) Clearly, there has been a slowdown in the reduction of undernourished in the world. […] The overall decline in the number of undernourished in the developing regions hides contrasting trends in different countries: only 32 of the 99 developing countries studied recorded a decrease in their numbers of undernourished between 1990-92 and 1997-99. The total reduction achieved by this group amounted to 116 million people. […] The decrease in the proportion of undernourished in these countries has not been sufficient to offset the effect of population growth. (*SOFI* 2001)

(7) A reduction in capital inflows will mean that consumption must be reduced. (*SOFI* 2009)

(8) Rapid population and labour-force expansion combined with inadequate domestic growth has stretched the capacity of the economy to absorb workers. (*SOFI* 2009)

In the quoted examples the extensive use of nominalization endows the texts to which they belong with great density and compactness since de-verbal nouns condense whole clauses and even sentences into nominal phrases.

As far as lexis is concerned, the tendency towards a domain-specific use of English is already prominent in the 2000 edition. Thus goes the caption to one of the maps interspersed within the text:

(9) The depth of hunger is measured by the average dietary energy deficit of undernourished people – not of the population as a whole – expressed in kilocalories per person per day. (*SOFI* 2000)

Phrases like 'body mass index', 'stunting', 'wasting', 'dietary energy supply', 'macronutrients', 'micronutrients' and many more are widely used in this issue, which clearly steers in the direction of technicalization of discourse. A great deal of technical terms are preferred to vaguer forms, yet all lexical items belonging to some specialized variety of the language are regularly defined or otherwise explained, as shown in the quote. The techniques used are those typical of scientific popularization, in particular denomination, definition and explanation (Calsamiglia/van Dijk 2004; Garzone 2006). Sometimes the explanation is provided by the use of synonymous phrases. An example can be found in the title of the section including the above mentioned map, which reads:

(10) Depth of hunger: how hungry are the hungry? (*SOFI* 2000)

The same phrase is coupled with an exact definition: "Depth of hunger: number of kilocalories missing from the diets of the undernourished," and in another part of the chapter it is paraphrased as: "The depth of hunger, or food deficit". The 'prevalence of hunger' is defined as the 'percentage of the population whose food intake falls below the minimum requirement', while 'malnutrition' is described as 'the entire range of problems that can occur when dietary energy and/or nutrient intake are insufficient, excessive or simply imbalanced.' The term is further analyzed into the variety of meanings it may cover:

(11) At one end of the malnutrition spectrum is the problem of undernourishment and undernutrition, often described in terms of macronutrients. Low dietary energy supply, wasting, stunting, underweight and low body mass index (BMI) are all used to identify the problem. At the other end of the spectrum is the problem of overnourishment, leading to overweight and obesity. A high BMI is one indicator of the problem. (*SOFI* 2000)

Macronutrients and micronutrients also receive accurate definitions:

(12) Macronutrients, the energy-providing food components (*SOFI* 2000)

and

(13) Micronutrients – minerals and vitamins – are needed for proper growth, development and function. (*SOFI* 2000)

Slightly more specialized medical terms such as 'iron deficiency anaemia', 'iodine deficiency disorders', 'vitamin A deficiency blindness', 'calcium deficiency', 'osteoporosis' and 'severe vitamin C deficiency' are incorporated in paragraphs which amply clarify their meaning. 'Body mass index (BMI)' is glossed as 'an anthropometric standard for defining the body composition of men and women' and further technical information is supplied in the form of a mathematical formula:

(14) A person's BMI score is calculated by dividing the individual's weight (in kilograms) by his or her height (in meters) squared: BMI = body weight (kg) / height2 (m). (*SOFI* 2000)

The 2000 *SOFI* report starts with a list of glossed acronyms, which facilitates the layman's decoding effort. Another list of acronyms opens the 2001 report, while later editions introduce technical terms from other specialized varieties of the language, invariably accompanied by explanatory phrases. Moreover, the 2000 and 2001 reports include a glossary, focusing on the definition of biomedical terms. The glossary added to the 2001 edition also includes words and phrases from economic discourse, like 'Dollar purchasing power parity' and 'Gini coefficient,' which are respectively defined as 'the purchasing power of a country's currency in relation to the US dollar; the number of units of a given currency required to purchase a basket of goods and services valued at US$1 in the United States,' and 'an aggregate numerical measure of income inequality ranging from 0 (perfect equality) to 1 (perfect inequality). The higher the value of the coefficient, the higher the inequality of income distribution; the lower the value, the more equitable the distribution of income.' Economic discourse plays a significant role in the post Millennium editions. In particular, the 2006, 2008 and 2009 reports abound in moderately specialized though transparent terms and glossed acronyms such as 'ODA' (Official Development Aid), 'GDP' (Gross Domestic Product), 'LIFDC' (Low-Income Food-Deficit Countries), 'MFIs' (Microfinance Institutions), 'FDI' (Foreign Direct Investment) and 'IMF' (International Monetary Fund). As regards the use of technicalities from other varieties of specialized discourse, the 2003 report introduces a number of terms stemming from meteorology and agricultural and water management technology, like 'erratic rains', 'high evaporation', 'flood irrigation', 'spate irrigation', 'stream flow diversion', 'groundwater recharge', 'runoff spreading systems', 'road runoff', 'periodic crop water deficits', 'in situ water conservation', 'eyebrow terraces', 'contour strips', 'furrows', 'contour bunds', 'trash lines', 'conservation tillage', 'subsurface dams', 'surface dams' and 'spring development'. A technical acronym mentioned in the same text is 'IPM', which is explained as integrated pest management. In this case as in other cases the 'hard words' (Gotti 2003) used in the reports are made accessible to the reading public through such devices as definition, paraphrase and explanation, which function as typical techniques for the popularization of specialized discourse (Garzone 2006). As a matter of fact, accessibility is one of the most prominent qualities of the *SOFI* reports, which are aimed at a double public, consisting in the community of practice originating the texts and in the international community made up of world citizens who are becoming more and more sensitive to such questions as justice and sustainability. All human beings are potential readers of the *State of Food Insecurity* reports.

4. Conclusions

This chapter has analyzed the *SOFI* reports from a diachronic perspective, identifying them as belonging to an interesting sub-genre within the area of institutional communication, namely that of the annual report drafted by an international organization. The evolution of the *State of Food Insecurity* report shows a tendency towards the technicalization of discourse, combined with an effort at explaining those terms which may otherwise sound impenetrable to the lay reader. A close analysis of the various issues reveals a double line of development: one points in the direction of specialization of discourse, while the other consists in a steady effort at enlarging the reading public both through a consistent use of popularizing strategies and through a wider circulation of the texts. The two trends have been examined in their textual and contextual implications.

From 1999 to 2009 the *SOFI* reports have seen the focus shift from a general definition of food insecurity and a cursory survey of the countries and populations affected by the phenomenon to a careful analysis of the circumstances originating its pervasiveness and depth. The titles of the various editions indicate this shift. 'Food Insecurity When People Live with Hunger and Fear Starvation' of the first four reports changes to 'Monitoring Progress towards the World Food Summit and Millennium Development Goals' of the fifth and sixth reports, which in turn switches to 'Eradicating World Hunger – Key to Achieving the Millennium Development Goals' of the seventh report, modified to 'Eradicating World Hunger – Taking Stock Ten Years after the World Food Summit' of the 2006 edition, while the 2008 issue bears the title 'High Food Prices and Food Security – Threats and Opportunities' and the 2009 edition concludes with 'Economic Crises – Impacts and Lessons Learned'. The macrostructures of meaning show that the State of Food Insecurity reports combine a fundamental concern with the multifarious and fluid reality of access to nourishing food and a keen interest in the opportunities for a more equitable distribution of world food resources. This combination of the two essential communicative purposes of informing and requiring action provides the key to the choice of text types which best suit the authors' intentions. An analysis of the titles of the reports reveals a moving balance between the provision of up-to-date information and the call for effective measures on the part of the international community. The working hypothesis underlying the research was that change in the context characterizing the drafting of the reports has influenced the handling of discursive practices and the distribution of text types, as can be noticed in the return to a massive presence of descriptive-narrative sections in the latest editions, published in times of severe economic and financial crisis which needed detailed analysis as a preliminary step to a renewed call for action.

In the *SOFI* reports, genre evolution largely coincides with a few significant changes in the pattern of communication which frames the various issues. The line of evolution emerging from that pattern is one of a progressive choralization of both encoder and decoder. The community of practice originating the reports becomes more and more inclusive and addresses a growing public. The double trend towards technologisation of discourse and the adoption of popularizing strategies reveals the drafters' wish to raise awareness among the readership, combined with a desire for greater accountability, which is also connected with the availability of the reports on the Internet and the consequent increase in the participation framework.

A more detailed textual analysis of the reports focusing on the categories of modality and transitivity – which lies beyond the limited scope of this chapter – may prove extremely useful for a deeper understanding of the genre, which would also benefit from an in-depth study of the move structure of the reports as well as of the strategic handling of genre-mixing and embedding.

CRISTINA PENNAROLA / GERMANA D'ACQUISTO

The Genre Evolution of the UN Secretary-General's Annual Reports from 1953 to 2010*

1. Introduction

The major influence of the United Nations on current affairs has been extensively examined from different perspectives (historical, political, military, organizational, sociological), but relatively few studies have dealt with the rhetorical aspects and ideological implications of its discursive productions (Rist 2002). In particular, no specific attention has yet been adequately paid to the UN Secretary-General's annual reports, notwithstanding the significant role they play in disseminating information on current affairs and orienting international policies. Rist (2002) examines the folktale narrative structure and formulaic style of all UN reports drawing attention to the simplifying evil and good forces dichotomy as well as to their undefined actors – e.g. 'developing countries', 'stakeholders', 'donors' – and identifies these categories as the hallmarks of the universal and abstract quality of UN discourse. In a similar vein, Senarclens (2002) highlights the UN's idealistic agenda and language, while Maingueneau (2002) examines the UN reports as an example of self-legitimating discourse that relies on a neutral language code ignoring or disregarding ethnic and cultural specificities. Bellier (2005) complements Maingueneau's linguistic analysis with an anthropological approach founded on real-life obser-

* This article was planned jointly by the authors. In particular, Cristina Pennarola is responsible for sections 1, 2, 3, 5, 7 and Germana D'Acquisto is responsible for sections 4 and 6. Germana D'Acquisto is also responsible for the quantitative analysis carried out by means of Wordsmith Tools. This study is part of the National Research Programme *Tension and Change in English Domain-specific Genres* (prot. 2007JCY9Y9_004).

vation that seeks to anchor the very notions of discursive community and 'modes of enunciation' to the concrete communicative situation and the specific deliberative procedure adopted at UN meetings. However, despite focusing on critical discursive features, none of the studies mentioned has specifically examined the UN Secretary-General's annual reports in a diachronic generic perspective.

1.1. Methodology and corpus

In recent times there has been a tendency in genre analysis to shift attention from the intrinsic structural properties of texts to the interactional and negotiated aspects of community discourse (Swales 1990; Bhatia 1993 and 2004). Unlike texts, which can be analyzed in isolation as self-contained units of language and discourse, genres are primarily defined by their purpose and their relationship to the discourse community taking responsibility for their adherence to prototypical content, positioning and form (Swales 1990: 52). The relation between purpose and textualisation appears to shape the typical features of genre at the interface between discourse, concerned with social and cultural issues, and language, focused on lexico-grammatical features, patterns of textualisation, and 'moves' (Hüttner *et al.* 2009).

This chapter examines the discursive and linguistic features of the UN Secretary-General's annual reports over the last 60 years with the aim of highlighting the evolving features of this genre and demonstrating its significant development from an informative genre addressing the restricted discourse community of the UN General Assembly into a self-promotional genre aimed at the public at large. This study is based on a corpus of 37 UN Secretary-General's Annual Reports on the Work of the Organization from 1953 to 2010 made available in a digital format by the United Nations Dag Hammarskjold Library. The coverage is uneven as many of the early documents are still in the process of being digitalized; for this reason, only five reports in this corpus relate to the first two decades of the UN, while the last thirty years from the eighties to 2010 are fully documented. However, as it was possible to identify linguistic and structural similarities across the early reports despite the long intervals between them, their limited number has been considered uninfluential in the analysis and interpretation of the findings.

The reports were analysed according to a genre analysis framework (Swales 1990, 2004; Bhatia 1993, 2004) taking into account their main generic features: structure and 'moves', communicative purpose and targeted discourse community (Swales 1990: 58). In addition, some lexico-grammatical features such as the use of pronouns, tense and the modal auxiliaries *will* and *would* were selected for analysis as consistent variation in their use across the whole corpus has highlighted sharp differences in textual encoding between the early and the recent reports. Their analysis has contributed to clarifying some aspects of the discursive process of UN reporting as it has evolved over time from a factual account of political dynamics into a grand narrative of past and present challenges.

On the basis of their structural characteristics, the UN reports were grouped into three periods: the early reports from 1953 to 1976,[1] the transitional reports from 1977 to 1991, and the 'new' reports from 1992 to the present day, testifying to a 'new world order'[2] and to the advent of globalization in a closely interconnected, borderless society.

Our study is mainly focused on the early and 'new' reports as these two groups represent the extremes of a cline, as shown by the analysis of their marked differences. While the first reports are very lengthy and give detailed accounts of all the major and minor events and meetings taking place in the year surveyed, the 'new' reports are much shorter narratives where events are condensed into a well-designed story line. The transitional reports resemble the 'new' ones on account of their condensed narration as they also tend to synthesize the main events and policies. However, they are much shorter – around 10 pages compared to the 70 pages of the recent ones – and also lack the use of stylish visuals and quantitative data typical of the

1 The early UN Secretary-General's annual reports date back to 1953, 1960, 1961, 1975 and 1976.

2 According to President George Bush, who first used this phrase in his speech to Congress on March 6th 1991, the 'New World Order' entailed "the sovereign state as the key unit of international relations; respect for the norms of non-aggression and non-intervention; and support for international law and institutions" (Brown/Ainley 2005: 198).

most recent reports. Moreover, unlike the early and 'new' reports, they only superficially deal with global problems as they advocate the intervention of the international community rather than illustrate the concrete step-by-step initiatives in support of human rights and economic development.

2. The genre(s) of UN reports

The documentation produced by the United Nations is massive, covering all aspects of UN policy-making, in addition to administrative and bureaucratic matters, particularly because all its main organs, and in turn their subsidiary bodies, functional commissions, specialized agencies, Departments and Offices have to keep a record of their activities, and are asked to prepare reports, surveys or special studies on specific events or subjects, ranging from technology to delinquency, from aid relief to national debt (Department of Conference Services 1984: V). This vast documentation output, closely monitored by the Editorial, Terminology and Reference Service and by an international team of editors, has a largely informative purpose consisting of reports and other factual pieces of writing. Depending on the addresser and addressee, various kinds of UN reports can be identified: research reports of the Secretary-General; internal department reports; mission reports; public information reports; technical reports; conference or meeting reports; inter-agency reports; incident reports; analytical reports on application of rules and policies (United Nations 2007).

Despite the variety of sub-genres, UN reports often have similar purposes, namely to inform UN staff, or non-specialist readers of a situation, problem, procedure or policy. In addition to the informative function, they may also perform a persuasive function, laying out recommendations or formulating requests. However, despite the basic informative core common to all UN reports and the overarching stylistic concern with clarity, their political significance changes dramatically as a function of the communicative scope, i.e. whether they directly address the Member States' representatives (or other equally influential recipients), and have wide public visibility, or whether they are simply meant for internal circulation and for recording purposes.

Of all the reports authored by the Secretary-General, the *Annual Reports on the State of the Organization* are the most significant and influential for a number of reasons: first of all, they survey the main policies implemented by the United Nations; secondly, they illustrate the future lines of action agreed upon by the international community; and lastly they respond to the demands for greater accountability increasingly placed on the United Nations as on all large organizations (Grant/Keohane 2005): in the words of former Secretary-General Boutros Boutros-Ghali, the UN Secretary-General's annual report "is itself an effort at transparency, revealing both the strengths and weaknesses of the organization to the widest possible audience" (*Report* 1995: 131).[3]

3. The early reports

Since its birth in 1945, the United Nations has embodied a dream of justice and welfare across borders, acting as policymaker and disseminator of good practices (Brown/Ainley 2005: 136-137) in a world characterized by fast-paced sociopolitical changes – decolonization; the end of cold war; the strengthening of an Islamic coalition; the outbreak of the so-called 'new' interethnic wars (Kaldor 1999), and the growing threat of international terrorism. However, one of the changes which has most affected the United Nations governance has been the spread of ICT on account of the powerful influence of the new media on politics and of the steady democratization of the decision-making process allowing "individuals and non-state organizations to influence the expanding international human rights regime" (Pubantz 2005: 1291). Before the start of the UN Bibliographic Information System and Digital

3 In this chapter each of the UN Secretary-General's annual *Reports on the State of the Organization* is referred to as *Report* followed by the year of publication.

Document System in 1993, the official documents of the United Nations – even the most influential – were only available in print and destined for a very restricted audience composed of UN staff, politicians and diplomats; by contrast, today, the largest part of UN documents is available even to the general public in electronic format.[4] This change in communicative situation (channel and readership) has widened the scope and reach of UN documents and combined their administrative purpose with a broader informative and persuasive intent.

In this section, attention is placed on the text organization and structure of the early UN reports with the aim of illustrating their character of meticulous chronological accounts of the UN General Assembly discussions. In fact, the detailed chronological narration of the debates, evidenced by the prominence of verbal 'moves', frequency of indirect speech and use of past tense, sets the early reports apart from the recent ones which, by contrast, summarise the main action plans, and foreground actual or expected results ignoring the debates underlying UN policies.

3.1. Chronological sequencing

The early reports are voluminous documents of two hundred pages approximately, organized into dense one-block chapters in which all the topics under the attention of the General Assembly – from the admission of new members to the settlement of inter-state disputes; from violation of human rights to administrative and budgetary questions – are closely scrutinized. Even the lightest steps leading up to any event or decision are meticulously reported in a dense chronological sequence, as illustrated by the following excerpt on the India-Pakistan question:

(1) As indicated in the last annual report of the Secretary-General, the United Nations Representative for India and Pakistan informed the Security Council on *29 May 1952* that negotiations with the parties had been renewed. On *31 July*, he further informed the Council that the Governments of India and Pakistan

---

4 Since January 2010, printed copies of UN documents are no longer distributed within the Organization, something which should help protect the environment and save seven tons of paper per month, according to official sources.

> had agreed to a meeting at a ministerial level to be held in Geneva under his auspices. On *16 September*, the United Nations Representative submitted his fourth report to the Security Council, covering the negotiations carried out from *29 May* to *16 July 1952* in New York and the conference held in Geneva from *26 August* to *10 September 1952*. Summarizing the conversations and meetings with the parties, he stated that, on *3 September*, it had appeared that no agreement could be secured [...] (*Report* 1953: 29)[5]

The numerous time references supplied in the paragraph provide detailed information on the whole lengthy negotiating process that the UN representative for India and Pakistan is called to report by the General Assembly; thus, rather than succinct reports of the final results achieved in the negotiation, the UN Secretary-General's annual reports can be regarded as extensive accounts in which all the step-by-step discussions, meetings and proposals are thoroughly covered despite the little success they may have achieved. This articulate temporal sequence – which is repeated over and over again for every single item on the UN annual agenda – is not simply evidence of the extreme meticulousness of the early reports, but also of the attention placed on the decision-making process.

3.2. Reporting 'moves'

The early reports present a cyclical structure revolving around three phases that recur throughout the report: the initial proposal for a political, economic or social change that is submitted to the General Assembly or equivalent authority; the discussion of the proposal that generally extends over several General Assembly sessions or other official meetings; and the outcome decided by the ballot and consisting in a draft resolution:

(2) *[Proposal]*
On 12 August 1952, the representative of the United States of America introduced in the Commission, on behalf of his Government and those of France and of the United Kingdom, a supplement to the three-Power working paper

---

5 Italics are used in this example and the following ones to highlight significant words and phrases.

setting forth proposals for fixing numerical limitations of all armed forces. (*Report* 1953: 1)

(3) *[Discussion]*
In the course of the discussion, questions were asked concerning the effect of the proposed five-Power and regional conferences on the status and functions of the Disarmament Commission. (*Report* 1953: 2)

(4) *[Resolution]*
The draft resolution was voted on paragraph by paragraph, the only negative votes cast being in connection with the clause reaffirming resolution 502 (VI). The resolution as a whole, as amended, was adopted by 52 votes to 5, with 3 abstentions. (*Report* 1953: 5)

In terms of 'moves', i.e. segments of text classified according to their prototypical communicative purpose and the rhetorical action they perform (Swales 1990; Flowerdew 2005), the early reports mainly consist of reporting moves characterized by a prominent verbal core: describing current situations; voicing major concerns; raising objections; expressing criticism; presenting proposals and counterproposals; outlining amendments; summarizing decisions. The focus of these reports is placed on the negotiations unfolding rather than on the results achieved, as shown by the frequency of verbal processes throughout the reports:

(5) The United States representative categorically *denied the charges*, and *declared* [...]

(6) In presenting his *draft resolution*, the representative of Mexico *emphasized* [...]

(7) The representative of Ceylon *expressed the view* [...]

(8) During the *debate*, many representatives *spoke against* [...]

(9) In the *discussion*, the representative of Cuba *reaffirmed* his Government's right to *appeal* [...]

(10) *Referring* to the *negotiations mentioned* by the representative of Saudi Arabia [...]

All the verbal actions in the quotations above may be regarded as performing an argumentative move, in line with the pragmatic core of diplomatic negotiations.

3.3. Reported speech and indirect projection

The techniques of speech presentation have been extensively examined over the last few decades starting with Genette's seminal work (1972) and complex categories of narrative analysis have been applied to literary and non-literary texts (Leech/Short 2007; Rimmon-Kenan 1983; Toolan 1988). For the purposes of the present analysis, however, only the distinction between reported speech and implied projected speech (more commonly labelled 'free indirect speech') has been considered relevant and apt to highlight the early reports' major focus on the verbal events and speech acts occurring in the General Assembly sessions.

The kind of narrative prominently used in the UN Secretary-General's reports of the early period is projection, which "can be loosely defined as language representing the use of language" or in other words "utterances from a former occasion" (Young 1988: 25). A projecting verb such as *state, inform, argue, recall* etc., associated with UN representatives, usually introduces an indirect speech clause. However, given the lengthy nature of the accounts, the projecting phrase is very often omitted to avoid stylistically awkward repetitions. As recommended in the online writing manual for the United Nations, "In a summary of a speech or discussion, the reporting clause must be added initially to establish the pattern of indirect speech but should not be included in every subsequent sentence" (United Nations 2007). The blend of direct narration and implied projection may be illustrated by the following excerpt:

(11) On 19 March, the USSR representative, recalling the various proposals relating to the prohibition of atomic weapons and the reduction of armaments which had been made by the USSR beginning in 1946, stated that the absence of progress in the Disarmament Commission was due to the efforts of the United States, the United Kingdom and France to divert the Commission from its terms of reference and to prevent it from coming to any definite decisions. *Accordingly, the two questions of the United States representative were merely a trick.* (*Report* 1953: 4)

The last sentence, highlighted in italics, is an instance of projected speech inasmuch as the negative comment on the US representative's questions can only be interpreted as being the USSR representative's evaluation rather than a matter-of-fact description, which would blatantly clash with the principle of objectivity informing the United Nations code of conduct as well as the factual writing entrusted to UN officers. A distinctive feature of the early reports is that all the representatives' statements are recorded even when they are potentially disruptive, or when they are eventually withdrawn or disregarded by the Committee in the final decision, as shown in the following excerpt:

(12) In a letter dated 15 October 1952 addressed to the President of the General Assembly, the representative of Czechoslovakia requested that the following question should be included in the agenda of the seventh of the General Assembly: "Interference of the United States of America in the internal affairs of other States as manifested by the organization on the part of the Government of the United States of America of subversive and espionage activities against the Union of Soviet Socialist Republics, the People's Republic of China, the Czechoslovac Republic and other Peoples' Democracies". On 16 October, the General Assembly decided to include the question in the agenda of the seventh session. (*Report* 1953: 50)

The complaint voiced by the representative of Czechoslovakia is devoted a whole section in the 1953 UN Secretary-General's report, which provides full coverage of the eight meetings in which Czechoslovakia argued the case for a draft resolution condemning the United States for their acts of aggression in the internal affairs of other States. Although, eventually, the Czechoslovak draft resolution is rejected twice, all the claims, statements, counterstatements, and argumentations are meticulously recorded, and so are all the representatives' turns of speech. This attention to diplomatic negotiation is a distinct feature of the early reports, so much so that they can be regarded as the minutes of all the year-round meetings rather than as a synthetic account of the UN's main activities and actions.

4. The 'new' reports: structure and 'moves'

The new reports present a very articulate structure which consists of chapters including headings and subheadings, numbered paragraphs and numbered subparagraphs. This orderly organization of text may reflect the need to clarify the points covered and detail the results obtained by the Organization. Moreover, the presence of attractive visual aids such as tables and graphs complements the information provided in the reports. Despite the articulate formal structure, however, the narration tends to simplify the main actions and synthesize them regardless of their chronology.

The titles of the introductions – e.g. 'An opportunity regained', 'Renewal and reform', 'Contours of the new era', 'Facing the humanitarian challenge' – place the stress on the need to renew and reform the Organization, outlining the leading role of the United Nations in the global era. The structure of the 'new' reports is based on 'problem-solving moves': problems are highlighted, action plans are surveyed and recommendations are adopted taking into account the new leading role of the United Nations:

(13) *[Identification of problem]*
The situation of no two countries is precisely alike in this respect but, broadly speaking, and making due allowance for cultural differences, good governance comprises the rule of law, effective state institutions, transparency and accountability in the management of public affairs, respect for human rights, and the meaningful participation of all citizens in the political processes of their countries and in decisions affecting their lives. (*Report* 1997: 5)

(14) *[Survey of action plans]*
Nevertheless, such practices as the sale of children, child prostitution and child pornography remain critical problem areas and are addressed by a Special Rapporteur. The implementation of the Plan of Action on the rights of the child, recently developed by the High Commissioner, is an essential next step. (*Report* 1997: 6)

(15) *[Recommendations]*
Ensuring that humanitarian strategies as well as longer-term development aims are fully integrated into the overall peacekeeping effort is another main priority in the year ahead. These are the challenges we must overcome as we seek to adapt the essential implement of United Nations peacekeeping to the crises we face today and will face tomorrow, [...]. (*Report* 1997: 17)

The attention is placed on the actions performed rather than on the verbal procedures leading to those actions; as outlined in *A Guide to Writ-*

ing for the United Nations (Department of Conference Services 1984: 7), "A United Nations report should not say everything that can be said on a given subject, but should select and state those facts, and only those facts, that are necessary to understanding of the subject for a given purpose". The meetings and written documents – conferences, treaties and conventions – originating UN actions are then simply mentioned, while the emphasis is laid on the achievement of positive results:

(16) In the past 12 months, significant progress in arms regulation and disarmament has been achieved through multilateral treaty-making, in several cases through the Conference on Disarmament. Treaties covering a wide range of weapons of mass destruction have been negotiated, extended or bolstered, or have entered into force. The new, institutionalized review process of the Treaty on the Non-Proliferation of Nuclear Weapons commenced, the historic Comprehensive Nuclear-Test-Ban Treaty was concluded, the Chemical Weapons Convention came into force, and agreement was reached by the parties to the Biological Weapons Convention to intensify their efforts to negotiate verification provisions. (*Report* 1997: 13-14)

The different use of tense that can be observed across the early and the 'new' reports can be interpreted in the light of the different narrative focus. The analysis has shown that the early reports are distant narratives set in the past tense and past perfect, and their main function is to keep a record of all the negotiation steps and to sum up the stages of a question examined; by contrast, the 'new' reports, which focus on the ongoing progress towards "the challenges ahead", are presented as contemporary narratives set in the present tense and often projected into the future.

5. Viewpoint in the early and 'new' reports

As argued by Simpson (1993), the 'angle of telling' or the medium through which the events of any account are focalized may provide significant information about the ideological positioning encoded in that account together with the cognitive and emotional orientation of the focalizer. Although many factors like deixis, modality, and transitivity come into play when constructing a textual world, the use of personal pronouns is most revealing of authorial stance and the process of narrativization. In the genre of the UN Reports, for example, the shift from an objective narration to a first-person account seems to reflect the personalization – or, to some extent, customization – of institutional discourses that has taken place in the last few decades (Pubantz 2005; cf. Kress 2010: 20).

In parallel with the different communicative purpose and narrative focus of the early and the 'new' reports, the authorial stance significantly differs across the two sub-corpora. The early reports are characterized by an objective, impersonal narration in which all the actors take responsibility for their speech acts and an apparently omniscient narrator – very close to the figure of a recording secretary – accounts for all the verbal events and the actions stemming from those verbal events, whatever their scope and significance within the UN political agenda. By contrast, the 'new' reports are substantially first-person narratives in which the Secretary-General himself takes full responsibility for the policies and action plans implemented. The authorial stance is made evident by highly frequent self-references, but still the new reports show "the accurate, clear, concise, consistent writing which is the kind most desirable in an international organization." (Department of Conference Services 1984: 5):

(17) 77. The level of international cooperation in disarmament remains disappointingly low. This is especially evident in the Conference on Disarmament — a crucial part of the multilateral disarmament machinery — where in 2001 no consensus on a programme of work could be reached. I hope that the appointment by the Conference of three special coordinators will help to bring about some forward movement next year. Several multilateral agreements still await either entry into force or effective implementation. (*Report* 2001: 13)

The kind of evaluative judgment and commitment to a better future expressed by the Secretary-General since the nineties makes the 'new' reports closer to a personal narrative revolving around his decisive influence, in line with the leadership role assigned to this figure at the turn of the century (Skjelsbaik 1991; Johnstone 2003). The early reports, on the other hand, far from emphasizing the leading role of the

UN Secretary-General, tend to refer to the Organization, and to the various organs and committees as the UN policy-makers:

(18) Since the work of the United Nations and the specialized agencies has a universal character, the Department of Public Information makes special efforts to underline on every possible occasion the need for an understanding and a knowledge of their common aims and activities in the international field. (*Report* 1953: 154)

The Secretary-General's assignments – preparing reports and preliminary studies; submitting proposals; responding to requests; organizing meetings; convening conferences; surveying governments' policies, etc. – are described in detail, yet his actions often seem to respond to some official request rather than being the outcome of his own judgment and initiative:

(19) The General Assembly, after examining this documentation, adopted and proclaimed the Declaration of the Rights of the Child, consisting of a preamble and ten principles. It recommended that Governments of Member States, the specialized agencies concerned, and the appropriate non-governmental organizations, should publicize the text of the Declaration as widely as possible, and requested the Secretary-General to have the Declaration widely disseminated and, to that end, to use every means at his disposal to publish and distribute texts in all languages possible (resolutions 1386 (XIV) and 1387 (XIV) of 20 November 1959). (*Report* 1960: 154)

The very fact that the Secretary-General is consistently referred to in the third person – as any of the other actors in the Organization – contributes to downplaying his central role as the official author of the annual report and as the leader of the United Nations.

In the early reports the addresser is concealed in an impersonal narration and the addressees are never mentioned; by contrast, in the 'new' reports the frequent references to the Secretary-General as the sender of the message and initiator of the actions, together with his direct appeal to the 'stakeholders' and civil society, convey a strong sense of community. The marked differences between the frequencies of first-person pronouns across the two sub-corpora, as shown in Table 1, appear consistent with the objective stance of the early reports as well as the involved narration of the 'new' reports.

	EARLY REPORTS (703,588 WS)		'NEW' REPORTS (291,372 WS)	
	Frequency	%	*Frequency*	%
I	19	<0.01	932	0.32
WE	9	<0.01	321	0.11

Table 1. Frequencies for 'I' and 'We' in the early and 'new' reports.

6. Agent-oriented modality in the early and 'new' reports

Since English modality is a very complex issue, "inextricably embedded in contexts of social interaction", and covering a wide range of semantic nuances – jussive, desiderative, intentive, hypothetical, potential, obligative, dubitative, hortatory, exclamative (Bybee/Fleischman 1995: 2), the focus in this section is placed on the most frequent modals across the corpus. In fact, variation in their use may signal an evolving function in the genre of the UN reports.

Bybee and Fleischman (1995: 18-53) distinguish between agent-oriented modality, referred to the completion of an action and speaker-oriented modality, through which a speaker attempts to move an addressee to action. For example, the sentence "This commitment must now be translated into effective action [...]" may be regarded as an instance of agent-oriented modality since the authorial voice expresses a clear recommendation directed at the Member States, while "As we look towards days filled with promise, we must also be prepared to deal with uncertainty. [...]" seems to convey a speaker-oriented modality.

Modal	*Early (703,588 ws) %*	*'New' (291,372 ws) %*
MUST	0.02	0.08
CAN	<0.01	0.09
WILL	0.08	0.27
WOULD	0.22	0.04
SHALL	0.01	<0.01
SHOULD	0.20	0.05

Table 2. Frequency of modal auxiliaries in the early and 'new' reports.

The analysis of the modal auxiliary verbs, supported by WordSmith Tools 3.1 software (Scott 1999b), suggests that an agent-oriented modality is dominant in the whole corpus, conveyed by narrative *would* and *should* in the early reports and by commissive *will* in the 'new' reports. The value associated with either modal is in fact commitment to action, also meant as obligation, necessity or desire:

(20) [..] I *will* continue to devote every· effort towards […]

(21) […] Here, I *will* simply highlight three that are worthy of considering […]

(22) […] we *will* be able to overcome this difficult period .

(23) […] our multilateral institutional architecture *will* also require adapting our multilateral mechanisms […]

In the early reports *should*, which can often be found in reported clauses, is used to express the recommendations and proposals advanced by UN representatives:

(24) […] The President made an appeal that no action *should* be taken that might aggravate the very dangerous situation […]

(25) […] The Korean people, who have suffered so grievously, deserve and *should* receive this help from the United Nations.[…]

(26) […] The Committee recommended that the Assembly *should* adopt a resolution approving the report […]

(27) […] They expressed the view, however, that the General Assembly *should* not take any action on the French communication […]

Although in subordinate *that*-clauses, *should* has a "quasi subjunctive" meaning which makes it semantically empty, the clause that precedes it may convey weak obligation, as pointed out by Coates (1983: 67-68).

Together with *should*, the most frequent modal auxiliary in the early reports is *would* corresponding to the past tense of *will* in reported speech, in line with the prominent diegetic function of the early reports:

(28) […] and had also given assurances that it *would* not give clearance to any aircraft […]

(29) […] was further authorized to say that his Government *would* give the United Nations the utmost cooperation […]

(30) […] Soviet Socialist Republics, declared that they *would* support the Burmese draft […].

(31) […] resolution by which the General Assembly *would* deplore the presence of foreign forces in Burma […] […]

7. Concluding remarks

The analysis has shown significant linguistic and textual changes across the early and the 'new' reports: in the early reports, the meticulous narration arranged chronologically almost day by day reports on the slightest action or events, while in the 'new' reports only the most significant events are organized in a 'grand narrative' that can be regarded as a tribute to the principles and values of the United Nations. The focus is then on the results achieved rather than on the negotiating phases leading up to those results, as the prominence of problem-solving 'moves' appears to suggest.

In parallel with the different narrative focus, the authorial stance differs across the early and recent reports: in the former, an invisible and impersonal narrator records all the statements and action plans including those by the Secretary-General himself, who is consistently referred to in the third person; in the latter, by contrast, the Secretary-General takes full responsibility for all his policies and action plans with a first-person narrative. The analysis has shown that while the purpose of the early reports is to provide a record for the UN administration, the 'new' reports seek to involve the audience at large and stimulate their participation in global governance so as to contribute to spread economic development, education and healthcare at a global level (Nye 2008).

The analysis of the most frequent modal auxiliaries suggests a dominant agent-oriented modality conveying commitment to the UN policies across the two sub-corpora. The high frequency of commissive *will* in the 'new' reports, points to the urgent need to involve civil society in "building a better world" in the new global era, while the prominence of diegetic *would* in the early reports has a merely constative function.

The evolution of the UN annual reports from an informative genre aimed at the restricted United Nations readership to a self-promotional genre meant to engage and captivate the public at large can be explained as the reflection of major sociopolitical changes: the impact of the Internet and the new technologies has made the vast UN documentation available to non-specialist readership outside the United Nations and also stressed the need for greater accountability (cf. Chesterman 2004: 152-153). As former UN Secretary-General Kofi Annan underlines, "Better governance means greater participation, coupled with accountability. Therefore, the international public domain – including the United Nations – must be opened up further to the participation of the many actors whose contributions are essential to managing the path of globalization." (United Nations 2000: 8).

VANDA POLESE / STEFANIA D'AVANZO

From Making to Promoting Law: An Investigation into the Evolution of 'Food Safety' Discourse in EU Summaries*

1. Introduction

Genres are strongly related to specific purposes shared by a community; in Swales' words, "[...] except for a few interesting and exceptional cases, genres are communicative vehicles for the achievement of goals" (1990: 46). The rapid changes in communicative practices and patterns prompted by the global digital revolution have deeply affected the configuration of professional and disciplinary genres resulting in drastic changes or decline, while new genres have emerged and some existing genres have survived the impact of new technologies although in a re-mediated form (Sarangi/Polese/Caliendo 2011).

This study intends to investigate the institutional legal discourse of the European Union realised in the Summaries of European legislation[1] as a form of popularisation stepping from the specialised genre of Directives and provide a critical interpretation of the findings. More specifically, the research involves a diachronic analysis of EU Summaries vs. Directives with a special focus on the informative function of popularisation and the ideological implications detectable in the Summaries with regard to 'Food safety' discourse.

---

* This study is part of the MIUR-funded National Research Programme titled *Tension and Change in English Domain-specific Genres* (Prot. No. 2007JCY9Y9).
Although this study was planned and discussed jointly, Vanda Polese is responsible for Sections 1, 3, 4, 5.1, 5.2, 5.4 and 6; Stefania D'Avanzo is responsible for Sections 2, 5.3 and 5.5.

1 Online at <http://europa.eu/legislation_summaries/index_en.htm> (last accessed 29 July 2011).

The 'Summaries of EU legislation' are posted on a dedicated website. In the definition used to present them on the opening page of this website,[2] two main points appear especially worthy of notice: one is the attention given by the EU to the main function of Summaries as a vehicle for presenting "the main aspects of European Union (EU) legislation in a concise, easy-to-read and unbiased manner", and the other is a strong focus on the Institution (the EU) responsible for producing this popularised genre. The definition itself raises questions with regard to what should be included in or excluded from the Summaries, as decisions in this respect may involve political considerations and, as has been found in the Summaries, be primarily dependent on purpose rather than on audience. In order to meet their communicative purpose (i.e. to disseminate information about EU legislation to the general public), the Summaries, which are mainly informative and have no legal validity, are expected to redraft directives in a way that is typical of popularisation, i.e. without "alter[ing] the disciplinary content [...] as much as its language, which needs to be remodelled to suit a new target audience" (cf. Gotti 2003: 295). Nevertheless, the association between 'reader-friendliness' and 'unbiasedness', as the two main qualities of EU popularised legislation presented in the Summaries, needs to be investigated; in particular, issues of authority have to be raised, and the ideological implications they may have must be explored.

In light of the above, the aim of this study is (i) to provide a contextual framework to the analysis of the Summaries of EU legislation with a view to defining the language of popularisation vs. specialised language; (ii) to analyse the main features of the EU summary as a genre; (iii) to explore the value(s) of modal auxiliaries in the Summaries investigated, specifically *shall*, in association with forms of citation – i.e. simplified and non-simplified segments of specialised legal knowledge from the Directives – in an attempt to investigate issues of power

2 This is the exact definition: "The 'Summaries of EU legislation' website presents the main aspects of European Union (EU) legislation in a concise, easy-to-read and unbiased manner. It forms part of the Europa portal, which is published by the EU institutions." Available at: <http://europa.eu/legislation_summaries/index_en.htm> (last accessed 29 July 2011).

and control on the part of the EU in addressing a non-specialist audience with reference to food safety in a diachronic perspective.

It is to be noted that *shall,* a modal which is generally acknowledged to be typical of specialised legal genres, is retained in the Summaries (see also Polese/D'Avanzo forthcoming), and in particular in citations from Directives. Therefore, this study will endeavour to answer the following research questions:

- What function does the modal *shall* have in the citations featured in the Summaries? More specifically, does it retain the same value as it has in the Directives? If this is the case, what are the consequences of this discursive choice?
- Are the forms of citation used in the Summaries and, within them, recourse to modals typically associated with specialised legal discourse, in line with the purported function and nature of the Summaries as outlined on the 'Summaries of EU legislation' website?

2. Corpus and methodology

The corpus includes Summaries of EU Directives within the thematic area of 'Food safety' (1982-2008). For the purposes of this study the Summaries investigated have been subdivided into three subcorpora, published respectively in the 1980s, the 1990s, and the 2000s.

Studies of EU Directives aimed at harmonising Member States' efforts to grant refugees and displaced persons' civil and human rights and of Summaries of these Directives within the thematic area of 'Freedom, security and justice' have been previously conducted by the authors of the present chapter (see Polese/D'Avanzo 2010; D'Avanzo 2010; D'Avanzo 2011; Polese/D'Avanzo 2011). In these studies the focus was mainly on linguistic and legal vagueness, with a view to exploring the language employed in the representation of migrants and the ideological implications operationalising power roles. In the present study, the Directives on 'Food safety' are examined with a special focus on the features of the dissemination-oriented, simplified language instantiated in the Summaries.

The methodology employed is mainly based on Genre analysis (Whitley 1985; Hilgartner 1990; Swales 1990, 2004; Bhatia 1993, 2004; Gotti 2003), Critical Discourse Analysis (Kress/Hodge 1979; Fairclough 1989, 1995, 2003) and studies on modality (Hoye 1997, 2009; Palmer 2001; Williams 2005, 2006). The text interrogation software AntConc 3.2.1 has proved useful for providing quantitative data for the investigation of the co-text of specific words and phrases.

3. EU legislation on 'Food safety': the socio-political-legal context

The three decades under examination show a growing concern on the part of the EU, with a peak in the 1990s, for problems in the area of food safety. These were also prompted by international and social changes which had appeared on the global scene in the 1980s. Among these major world events, the 1986 Chernobyl nuclear disaster in the territory of the former Soviet Union had worldwide resonance and left the soil around the fall-out area contaminated by radioactivity for several years with effects being perceived as far away as Norway and the UK. Another turning point in the global setting was the geo-political shift caused by the fall of the Berlin Wall in 1989 as it opened up new scenarios for food through the spreading into Eastern Europe of fast-food outlets and across the world of big-name food brands going global, with products from across continents influencing international markets. The rapid changes in society in the 1980s resulted in a new way of life based on time-saving convenience and a quicker and busier life which brought about acceleration in the growth of pre-packaged and ready-to-cook foods as well as a growing preference for eating out. However, over-production problems rising in Europe in the 1980s urged a series of reforms within the EU's Common Agricultural Policy and stressed the necessity to prefer quality over quantity in food production also causing food companies to turn to health and ethical aspects in their marketing.

Nonetheless, many outbreaks of food-related diseases were reported by the media in this decade and resulted in food safety becoming an issue of growing public concern. The Italian cases of premature sexual development and accelerated growth in infants in consequence of baby foods containing a veal extract with high concentrations of a particular growth hormone, a number of Salmonella cases linked to eggs and dairy products consumed at public institutions in the UK and a virulent new type of *E. coli* bacteria identified as a food safety threat in 1982 called for EU specific legislation which set out measures to monitor and reduce the incidence of these diseases and to enforce the adoption of strict hygiene rules to reduce their occurrence through the food chain to the greatest possible extent. In 1986, further serious problems arose because of BSE – Bovine Spongiform Encephalopathy, also known as 'mad cow disease' – which was first diagnosed in cattle in the UK. The high number of BSE cases in the UK and Ireland caused the EU to place restrictions on exports of live cattle and beef from the UK to other Member States and introduce a whole body of legislation to tackle BSE and protect consumers.

In consequence of facilitated imports of foodstuffs that could not be produced in great quantity in the EU and of strengthening trading relations with developing countries, in the 1980s the world grew more and more globalised in terms of food trade, with major global food brands being regularly selected by consumers in supermarkets and the number of fast food outlets growing across member States.

Accordingly, the following decade, the 1990s, saw arrangements in view of the free movement of goods, services, people and capital and harmonization of rules in order to ensure consistent high standards in the production, transport and sale of food across Europe independently of the place where products were made. In this decade, a revival in interest in home-made food was once again a recognised trend, also encouraged by the appearance of celebrity chefs on TV programmes, alongside the spread of fast-food outlets across Europe due to enthusiasm for ready-made meals and time-saving products. Moreover, food manufacturers found ways to obtain essential nutrients from new sources, and additional vitamins and minerals were added to many foodstuffs, like milk and breakfast cereals.

This notwithstanding, the 1990s was the decade of one major food crisis in Europe where BSE became a real problem. Since the general public grew deeply concerned about possible links between BSE and the feed given to cattle, and about how safe products were, the European Commission adopted strict measures to protect human and animal health (e.g. a ban in 1994 on feeding meat and bone meal (MBM) to cattle, sheep and goats, followed by a ban on processed animal protein being given to any animal farmed for food production as well as strict measures concerning animals slaughtered for human consumption) in order to detect, control and eradicate BSE, which actually began to decline from the mid-1990s and fell dramatically over the 2000s. An additional concern in the 1990s was the presence of dioxins (cancer causing compounds) which were found in animal feed in Belgium; many affected animal and food products had to be withdrawn from the market and destroyed and temporary restrictions had to be imposed on the trade in meat and dairy products.

A result of the effective approach to food safety and consumer protection adopted by the EU to regain confidence in the food being sold in Europe was the Food and Veterinary Office (FVO, 1997), aimed to ensure inspections and compliance with EU food safety and animal health rules, and the Directorate-General for Health and Consumer Protection (1999), designed "to help make Europe's citizens healthier, safer and more confident", including the development and implementation of EU food safety legislation.

Over the years, as food production became more complex, market choices grew wider and media coverage of food issues increased, consumers took greater interest in food products and gradually became more alert to issues related to food safety. The increased consumer awareness pressed the adoption of an innovative approach to provide reassurance regarding the safety, the quality and nutritional benefits of food products. The recommendations adopted by the EU through the *White Paper on Food Safety* (2000),[3] which set out a 'farm to fork' ap-

3 Available at <http://ec.europa.eu/food/food/intro/white_paper_en.htm> (last accessed 29 July 2011).

proach to ensure the highest possible standards of food through food safety controls, consumer information and imports, were made into a law in the EU's General Food Law (2002). This was meant to ensure that all food, feed and animals for food production would be traceable from 'farm to fork', with detailed records to be kept by anyone in the production, manufacturing and sale of food and feed about the origin and destination of food products, so that any food considered unsafe for public health could be detected and withdrawn from the market and consumers could take appropriate actions with unsafe products. Further important food safety rules followed on the wake, such as the creation of the European Food Safety Authority (EFSA) and the Rapid Alert System for Food and Feed (RASFF) in 2002 as well as the so-called 'Hygiene Package' (adopted: April 2004; enforced: 1st January 2006).

The 2000s are also marked by attention to novel food. Owing to globalization, new food substances and ingredients, used in non-European countries and thus unknown to Europeans, have been introduced to the EU market causing a steady growth in new types of food often produced using new technological procedures and whose traits need to be verified. Under EU legislation, such novel foods need to undergo a safety assessment before being placed on the market and are only allowed to be used or sold in the EU if they are shown to be safe and to raise no problem of food safety or human health.

4. Legal discourse: specialised vs. popularised genres

The world of specialised discourse shows disciplinary variation at various levels (lexical, morphosyntactic, textual, pragmatic) in such a way that common features of specialised discourse can be found to be associated with more specific disciplinary features.

The type of specialist-to-specialist communication which characterises the highest type of specialised discourse, i.e. when a specialist addresses other specialists in the same disciplinary field to discuss or present studies to a specialist audience, exists alongside a second type of specialised discourse, which occurs when specialists address non-specialists for educational purposes, and a third one, when specialists address a wider audience to provide information on specialised topics. In each of these cases, choices of lexis, meaning making strategies, and authorial selection of specialised materials are subservient to choice of audience and purpose.

EU Directives, as instances of a specialised legal genre, are normative in nature, requiring the intervention of Member States to fulfil the legal message in compliance with the norms provided. They are legally binding, and although they provide a mandatory deadline for national implementation within the hierarchical legislative territorial framework of Member States, they concede a margin of freedom to Member States in achieving the prescribed objective (see Caliendo/Di Martino/Venuti 2005; Polese 2006; Polese/D'Avanzo 2010; D'Avanzo 2010). Member States are therefore required to implement Directives by adapting the national ruling system, and to enact Directives by a given deadline or through stages within the deadline, and respecting the hierarchical legislative framework of the State territory.

Popularisation of legislative texts aims at making them accessible to a wide readership through easification devices which help the reader process the texts while preserving the originality and authenticity of the source genre. The complexity of the syntactic structure and the high density of information may create obstacles to a clear understanding of a legislative text by the lay reader. Thus, reducing the information load is a primary easification procedure also identified as 'textual mapping' (Bhatia 1993: 211). Other easification devices are: clarification of cognitive structuring, indicating legislative intentions and illustrating legislative issues.

The target audience and purpose set up a boundary between the two legal genres (summary and directive) as they account for decisions to be taken in the transformation process from a source genre/text (directive) into a target genre/text (summary) with regard to lexis and textual organisation.

Decisions need to be taken as to the selection of what aspects to simplify and what contents to include or exclude, as well as to what language to use, the summary being a simplified representation of specialised legal knowledge. Owing to this type of decisions, summaries may be subject to bias since specialists may recur to cutting and/or using linguistic devices designed to persuade a non-specialist audience to suit their purposes (cf. Hilgartner 1990: 519-520).

Furthermore, as in the translation process, in the popularisation process the product happens to be an imperfect equivalence of the source genre/text, being the result of a "translation from one system of discourse to another" (Whitley 1985: 7). Popularisation also implies simplification, urged by a need for clarity when addressing a non-specialist audience, a limited use of specialised lexis or technical terms and a preference for words drawn from general language (cf. Gotti 2003: 296-297), which are in part responsible for the loss of precision. As a redefinition of the message of the source text, popularisation may allow room for discretion in the production process for enacting forms of power which turn these representations into something other than politically neutral. This aspect has been investigated in the Summaries with regard to processes of hybridization. Hybridization has been defined as a kind of 'colonization' (Bhatia 2004: 58),

> [...] involving invasion of the integrity of one genre by another genre or genre convention, often leading to the creation of a hybrid form, which eventually shares some of its genre characteristics with the one that influenced it in the first place.

In the Summaries under examination, hybridization appears to be generated by the embedding of simplified and non-simplified segments of specialised legal language migrating from the source (Directive) into the target genre/text (Summary) mainly in association with *shall*-occurrences.

5. Citation and modality in the Summaries

5.1. Knowledge and features of discourse

As highlighted by Gotti (2003: 310), what characterises specialised discourse is "the particular semantic and pragmatic values it encodes, which in turn depend on the specificity of content and on knowledge shared by the interlocutors within a professional community". In complying with the norms that are explicitly or implicitly recognised within the specialist community, specialists leave no space to subjectivity in discourse. Two key points in discourse production and comprehension are the amount of shared knowledge of the participants and its being context-dependent. This means that substantial amounts of knowledge are relied on in understanding words and sentences, thus establishing local coherence and generating any kind of inference among other aspects of discourse understanding (cf. van Dijk 2005). Understanding the way knowledge is managed in discourse can thus help reveal the ways in which power roles are operationalised in communication and discourse through specific knowledge.

Another important aspect concerns the criteria that each community relies upon to distinguish beliefs shared as knowledge from others that are not, the former being "presupposed in the public discourses of that community" (van Dijk 2005: 73). Since context categories influence all structures of discourse (lexical, syntactic, rhetorical, speech acts, etc.), many aspects of discourse rely on assumptions as to degrees of shared/unshared knowledge. Thus, when it is assumed that a certain type of knowledge is shared knowledge, the speaker/writer may presuppose it, though explicitly or implicitly, or make explicit reference to it when it is not accessible. In the case of the Summaries, the source knowledge is provided through links which directly lead to the source text, but also through segments of source knowledge embedded in the target discourse which orientate the reader towards a certain interpretation stressing the legal value of the message and enhancing the role of the authority.

Moreover, owing to the nature of the communicative interaction, intentions, the audience, the institutional setting, relations among participants and other contextual properties, the type of knowledge which is being expressed or presupposed is not only dependent on what the recipients already know, but also on what text producers think it is interesting or relevant for them (cf. van Dijk 2005: 76) or on what text producers want recipients to know. The knowledge of language users about the knowledge of the recipient(s) is in fact crucial for the control of many important aspects of discourse, e.g. the type and amount of information which is explicitly expressed or reminded or presupposed.

In addition, the migration of segments of specialised legal discourse to popularised texts also constitutes a way to bridge the gap between the specialist and non-specialist community.

Popularisation requires that a specialised register with its own norms, patterns and style (which affects both terminology and reasoning through particular discourse practices) has to be 'translated' in order to be addressed to a wider non-specialist audience, thus demanding recontextualization and resemioticization both at a global and at a local level in the text. Our study mainly explores modality in association with 'citation' as discursive strategies which allow the drafter to preserve the original flavour of the specialised legal genre for specific communicative purposes.

5.2. Modality

As an inherent pragmatic phenomenon, modality involves the many ways in which attitudes (speaker stance) can be expressed towards the 'pure' reference-and-prediction content of an utterance (Verschueren 1999: 129). As a linguistic phenomenon, modality should be understood at the level of the utterance. It is particularly operative in processes of meaning making through language and can be understood in relation to the context of use where speakers act in accordance with their frames of mind and communicative needs and purposes. Context has a role in capturing modal usage and thus makes it possible to disambiguate ambiguity created by the polysemous nature of modals (cf. Culler 1988: xiv; Verschueren 1999: 75-114; Barbiers 2002: 1-7). Context provides a pragmatic framework to modals in relation to modality and the power dimensions involved (cf. Hoye 1997: 63-67). Thus, since language is ideologically loaded (Fairclough 1989), modal selection characterises the utterance in terms of speaker authority.

As to *shall*, this modal auxiliary has emerged as the most prominent modal in previous analyses of legislative texts (see Garzone 2001: 156; Caliendo 2004; Caliendo/Di Martino/Venuti 2005; Polese 2006, among others) in which the modal has been found to occur either with its performative or prescriptive value as a feature of legal documents. Of the two values – the performative, implying the contextual fulfilment of the action predicated, and the prescriptive, expressing the illocutionary force of an order where the addressee is required to act in compliance with the command – the latter has been found to predictably occur in the Directives and, quite unpredictably, in the Summaries on 'Food safety'. In the Summaries, in fact, no imposition of duty or obligation to a course of action can be expressed owing to the informative and non-legal nature of the genre; however, some explanation for *shall*-occurrences can be provided on the grounds of power relations with the total involvement of the authority encoded within such text types.

5.3. Values of shall *in the Summaries*

The findings resulting from the analysis of the frequency of *shall*-occurrences in the Summaries on 'Food safety' from 1982 to 2008 have been helpful to identify a role for the modal in the popularised genre. In EU Directives *shall*-occurrences convey deontic modality, which "involves the issues of directives and is associated with such notions as permission or obligation" (Lew 1997: 146), being "dependent on some kind of authority, often the speaker" (Palmer 2001: 70). The adoption of the modal auxiliary *shall* has for centuries been the most common way of imposing legal duties in prescriptive texts (cf. Williams 2005: 207). Its presence in this type of text "[…] seems to be accounted for by its original meaning as conveying obligation and thus expressing deontic modality which implicitly covers futurity" (Polese 2006: 103).

In the Summaries on 'Food safety', values of *shall* have been investigated taking into account the occurrence of the modal within phrases 'migrating' from the summarised Directives. Modality has thus been analysed from a wider perspective, more specifically, in terms of power relationships, that is, as "[...] a matter of the authority of one participant in relation to others" (Fairclough 1989: 105). The main assumption here is that the use of *shall* is aimed at providing legitimacy to the legal discourse and operationalising power relations in a popularised genre.

In order to analyse the role of *shall* in the Summaries on Food safety from a diachronic perspective, quantitative data are reported in the Table below:

	Shall
Eighties	*1.20*
Nineties	*1.43*
2000-2008	*0.26*

Table 1. Diachronic frequency of *shall* in the Summaries (pth words).

As is shown in Table 1, an increase in *shall*-occurrences can be found in the nineties. To interpret their value(s), *shall*-occurrences have been analysed in the Summaries focusing on the phenomenon of 'migrating quotes' as forms of 'citation' (see also D'Avanzo 2011) from the source text.

5.4. Citation

From a critical perspective, modality and citation are strategies adopted in popularisation not merely to convey information but also to influence readers' interpretation of the message. Identifying the role of citation allows us to understand whether quotes merely enhance credibility and attribution or rather strengthen orientation on the grounds of ideological implications.

Citation underlines what is regarded as especially important in the information provided (see Calsamiglia/López Ferrero 2003: 170) and is a marker of evaluation with the specific function of pointing to the topic of reference in the source discourse (SD). In line with Calsamiglia and López Ferrero (2003: 149), we adopt the critical perspective that "citation means managing the words of others to convey and serve the purpose of the writer" and expand it to identifying its main function as that of operationalising power roles by providing objectivity, attribution and legitimation. As a discursive strategy, citation orientates readers towards a certain interpretation of the message as part of the transaction process which is functional to linking the production process – aimed at construing the event – and the reception process – aiming at leading to a certain interpretation of the event (Charaudeau 1997: 72). Despite their respective distinctive traits, once on the Web, both the Directive and the Summary can be viewed in terms of media events since they share a wide mediated audience and conventions determined by the digital medium. In this context, citation can be found to play a specific role characterising the style of Summaries. However, none of the four styles identified by Calsamiglia and López Ferrero (2003) nor the ones identified by Politis and Kakavoulia (2006) (see Section 5.5) seem to fit in with the type of citation that has been found to play a role in the Summaries. A further type has been identified which has been called 'migrating quote' (see also D'Avanzo 2011). Two different variants of this type have been found to occur: the former, as unmodified segments of original knowledge directly migrating from Directives into Summaries with zero quotation markers and no use of reporting verb which may locate the agent/writer in a position towards the topic; the latter, undergoing easification processes (see Section 5.5). In both cases, they are embedded within the Summaries so as to be integrated with the co-textual items and hardly ever identifiable unless for *shall*-occurrences. This is a clearly identifiable marker of provenance from the Directives. The type of citation and the value and discourse role of the modal auxiliary can reveal relations of ideology and authority in the popularised genre with reference to EU 'Food safety' discourse. As a result, although no explicit interpretative cues are provided, interpretation is implicitly orientated in a way which enhances the role of the text producer/originator, which is the same both for Directives and for Summaries (i.e. the EU).

Although the Summaries as a genre can be included within the hypercolony of 'Reporting' genres (Bhatia 2004), they are characterised by no such devices as reporting verbs or a paraphrase of the source discourse (SD) within the target discourse (TD), but by features of formality – language with no affective involvement, more or less complex syntactic structures, impersonal and non-subjective use of language, zero quotation markers. Chunks from SD are reported in TD, resulting in the embedding of source segments of specialised knowledge in TD, pointing to a representation of the target discourse attributed to the 'authority' voice of SD through which the informative function is combined with orientation in the reception process. This is placed in an interpretational frame which is revealing of the institutional attitude towards the issues at stake in TD. TD is thus a vehicle for reproducing the main points in SD and above all a strategy for enhancing the legitimation of European legal provisions in a genre for the dissemination of specialised knowledge.

5.5. Simplified vs. non-simplified 'migrating quotes'

Investigation of the cotext of *shall* has revealed co-occurrence of *shall* embedded in 'migrating quotes' as forms of citation from SD to TD. The importance of forms of citation has been widely investigated in media and news discourse from a variety of perspectives. According to Bell (1991: 207), 'Direct Discourse' is the model par excellence in written news reports, whose function is to ensure faithfulness to a previously uttered speech or to provide factual information on socially important speech events. Calsamiglia and López Ferrero (2003: 155), instead, have analysed the phenomenon focusing on different types of 'citation'. More specifically, they have identified 'indirect citation' as one discourse, D1, with a single deictic centre, a subordinate clause introduced by a conjunction, and the correspondent agreement of tenses. 'Integrated citation', instead, has the form of indirect citation but with segments – of greater or lesser extension – being cited directly/literally with clear graphic or typographic marking (e.g. quotation marks or marked fonts, boldface or italics). Furthermore, 'inserted citation' consists of words brought into the main discourse by means of markers such as 'according to X', which have the function of assigning explicit words to a particular agent. Finally, 'direct citation' consists of the mixture of two types of discourse when "the two segments are connected through juxtaposition and they are signalled by a graphic marker such as (:)". The latter is similar to what has been identified by Politis and Kakavoulia (2006: 345) as 'direct quotation', defined as "the most explicit form of inclusion of a speech event in a discourse".

In the instances under examination in our study, instead, since no graphic or quotation markers can be observed in the Summaries when 'quotes' from legal discourse migrate from the discourse of the

Directives, the 'migrating quotes' are more similar to what Jucker (2006: 108) calls 'Free direct speech', which "lacks either the reporting clause or the inverted commas". More specifically, this type of speech is considered as a kind of Direct speech where "the characters apparently speak to us more immediately without the narrator as an intermediary" (Leech/Short 2007: 258).

Direct speech has been analysed as a phenomenon mainly occurring in news reporting and fiction to 'report' something 'told' by someone else in a more or less 'free' form. In the Summaries investigated, instead, 'migrating quotes' can be viewed as discourse devices originating hybridization of two legal genres and registers (Summary and Directive), i.e. specialised and popularised legal discourse. Co-occurrences of *shall* with hybrid chunks of text in this non-legal genre may be accounted for as an attempt on the part of the EU to preserve the legitimacy of legal discourse in a popularisation which is aimed at giving a "populist force to official voices" (Fairclough 1995: 71). For a closer investigation of the role of forms of citation in the Summaries, quantitative data are reported in Table 2:

	Migrating quotes	Simplified 'migrating quotes'
Eighties	0	1.20
Nineties	0.079	1.35
2000-2008	0.052	0.208

Table 2. Frequency of occurrences of '*migrating quotes*' vs. *simplified* '*migrating quotes*' in the Summaries.

As is shown in Table 2, a higher frequency of simplified 'migrating quotes' of specialised legal knowledge can be found in the Summaries compared to figures for 'migrating quotes'. As is also illustrated by the examples below, the fusion of the two registers seems to prevail (italics added in all examples):

(1) Member States *shall* take all measures to ensure that the substances and materials listed as extraction solvents in the Annex satisfy the following purity criteria:

(a) they *shall* not contain a toxicologically dangerous amount of any element or substance […] (Council Directive 88/344/EEC, Art. 3)

(1a) The Directive lays down purity criteria for extraction solvents, e.g. they *shall* not contain a toxicologically dangerous amount of any substance. (Summary of Council Directive 88/344/EEC, Art. 3)

In example (1a) from the 1980s subcorpus, omission of categorization can be observed (i.e. *substances and materials listed as extraction solvents* becomes *Extraction solvents*). Conversely, the segment *they shall not contain a toxicologically dangerous amount of any substance* appears to migrate from the source text, retaining *shall* – which is commonly found to occur in legal texts – despite the lack of doublets – another common feature of specialised legal texts (i.e. *Any element or substance* in example (1) turns into *any substance* in example (1a)).

Devices of this type can be found in the following instances from the 1990s and the 2000s subcorpora:

(2) In accordance with the procedure laid down in Article 23, the Commission *shall*, on the basis of the plans referred to in the second subparagraph, adopt the rules applicable to imports of animals for slaughter intended for local consumption and of breeding or production of animals in certain parts of the territories referred to in Annex I to Directive 90/675/EEC to take account of the natural constraints specific to these territories, including their remoteness from the mainland part of Community territory. (Council Directive 91/496/EEC, Art. 13)

(2a) The Commission *shall* adopt the rules which apply to imports of animals for slaughter intended for local consumption and the rules applicable to animals for breeding and production in certain parts of the territories in order to take account of the natural constraints particular to those territories and of their remoteness from the mainland part of the Community territory. (Summary of the Council Directive 91/496/EEC, Art. 13)

(3) However, if the Member States have not reached agreement within the time-limit referred to in Article 32(4) they *shall* forthwith refer the matter to the Agency, for referral to the Committee, for the application of the procedure laid down in Article 36. (Council Directive 2001/82/EC, Art. 32)

(3a) Where a Member State considers that alteration, suspension or withdrawal of an authorisation is necessary to protect human or animal health or the environment, it *shall* forthwith refer the matter to the Agency for application of the conciliation procedure described above. (Summary of the Council Directive 2001/82/EC, Art. 32)

In examples (2a) and (3a), in fact, forms of simplified 'migrating quotes' from SD can be noticed to occur in association with *shall* in the phrases *The Commission shall adopt the rules* [...] and *it shall forthwith refer the matter* [...]. Some summarising devices can also be noticed, such as omission of reference to further legislation (e.g. "on the basis of the plans referred to in the second subparagraph" in example (2)) and of parenthetic phrases (e.g. "for referral to the Committee" in example (3)).

The following instances are, instead, examples of 'migrating quotes':

(4) All approved semen collection centres shall be registered, each centre being given a veterinary registration number. *Each Member State shall send a list of semen collection centres and their veterinary registration numbers to the other Member States and to the Commission* [...] (Council Directive 90/429/EEC, Art. 5)

(4a) The Directive lays down that *each Member State shall send a list of semen collection centres and their veterinary registration numbers to the other Member States and to the Commission.* (Summary of Council Directive 90/429/EEC, Art. 5)

In example (4a) from the 1990s subcorpus, it is possible to notice entire pieces of legal discourse (in italics) directly quoted from the Directive without any kind of punctuation. The same strategy can be noticed in the following instances from the 2000s subcorpus:

(5) *The competent authorities of the Member States shall suspend or withdraw marketing authorization when it is clear that:*
(a) the veterinary medicinal product proves to be harmful under the conditions of use stated at the time of application for authorization or subsequently;
(b) the veterinary medicinal product does not have any therapeutic effect on the species of animal for which the treatment is intended [...] (Council Directive 2001/82/EC, Art. 83)

(5a) *The competent authorities of the Member States shall suspend or withdraw marketing authorisation when it is clear that:*
* the product proves to be very harmful;
* it lacks therapeutic efficacy [...]
(Summary of Council Directive 2001/82/EC, Art. 83)

In example (5a), the 'migrating quote' from the source text is located at the beginning of the textual fragment and is followed by a summarised schematisation as is common in the Summaries; it is a further instance of 'colonization' of the register of the Summaries by specialised legal discourse.

6. Concluding remarks

Further investigation will certainly add material to the interpretation of findings. As to the findings resulting from this study, the migration of specialised (simplified and non-simplified) segments with *shall*-occurrences from the Directives into the Summaries of EU 'Food safety' discourse may appear to be somehow unnecessary or redundant with regard to the informative and non-legal nature of the latter genre. However, such discursive strategies are functional to the conceptualisation of 'Food safety' discourse in EU legislation for dissemination, to a wide public, of special provisions and measures taken, particularly in the 1990s. The findings reveal an increase in *shall*-occurrences and in (simplified/non-simplified) 'migrating quotes' in the 1990s followed by a neat decrease in the 2000s; the increase recorded in the 1990s is all the more significant as 'migrating quotes' were totally absent in the previous decade.

Summing up, in the 1990s, a decade of great changes in Europeans' and the global world's eating habits, heightened consumer awareness and interest in food products in association with alarm to issues related to food safety, the discursive strategies described above appear to be functional to strengthening legitimation and credibility in EU's involvement in food safety, and providing reassurance on the safety, quality and nutritional benefits of food products in the global era.

PAOLA CATENACCIO

Genre-bending for Consensus Building: A Diachronic Perspective on Monsanto's Pledge Reports (2000-2009)

1. Aim and background

This chapter aims to investigate in a diachronic perspective the Pledge reports published by the biotechnology giant Monsanto over the 10-year period from 2000 to 2009. The reports, which are published yearly by the company to provide updated accounts of its sustainability-related initiatives, represent a specific instantiation of a fairly established, though comparatively new, genre of corporate communication: the sustainability/corporate responsibility report.[1]

By investigating these documents diachronically, the study aims to map the evolution of the discursive strategies deployed in the

1 By sustainability/responsibility reports I mean here corporate texts devoted to the "self-reporting of organizational socio-environmental interactions" (Spence/Gray 2007: 11). This type of reporting – which Spence and Gray call SER, or Social and Environmental Reporting – is "variously referred to as social accounting/reporting; environmental accounting/reporting; corporate social reporting; sustainability/sustainable development reporting" (Spence/Gray 2007: 12). The sheer variety of labels used to refer to one and the same type of discursive practice suggests that this is a domain which continues to be subject to variation, and where attempts at standardization have only been partially successful. However, thanks also to the effects of the ever more widespread adoption of GRI guidelines (Global Reporting Initiative, aimed specifically at establishing standards for sustainability reporting), over the last few years there have been signs that a recognizable genre structure is becoming established. A general outline of such structure has been proposed by Catenaccio (2011a), and will be referred to when testing the generic integrity of the Pledge reports *vis-à-vis* other instantiations of the genre .

reports over the last decade, thus contributing to the growing body of discourse- and language-oriented scholarly literature devoted to the exploration of corporate social responsibility (henceforth CSR). The assumption underlying this literature – which is shared in the present research – is that CSR is first and foremost a discursive construction. This should not be taken as a claim that there is nothing to CSR but rhetoric (a charge frequently laid at companies which are seen to invest heavily in CSR communication, but much less in actually improving their social or environmental performance): undoubtedly, the deployment of socially and environmentally responsible rhetoric is only warranted if it is matched by appropriate actions. However, insofar as CSR is part of a company's intangible assets (on a par with reputation, of which it is a component), it is essential that it is communicated, so that it can be leveraged on for reputation-enhancing purposes. In fact, CSR initiatives have often arisen in response to challenges which can only be described as rhetorical and indeed more specifically dialogic in nature. For instance, when companies such as GAP or Nike were charged by activist groups with exploiting child labour, they responded not only with policy actions, but also with massive communication campaigns aimed at restoring their reputation. But even when companies are not directly under attack, the need to establish a 'licence to operate'[2] requires the deployment of rhetorical strategies aimed at persuading a (presumably sceptical) audience that a company's activities are beneficial not only to its shareholders, but to society at large.

In light of the above, and building upon the findings of the above-mentioned growing body of literature (cf. Livesey 2001, 2002; Ellerup Nielsen/Thomsen 2007; Coupland 2005, 2006; Igalens 2006; Catenaccio/Degano 2011; Degano 2008), this study takes a discourse-analytical approach to CSR communication. Such approach relies on a

---

2 In the past, the notion of a business's 'licence to operate' referred only to a company's legal obligations. More recently, however, it has come to include – besides this legalistic requirement – also 'economic reality' requirements, such as the need to maximize shareholder return on investment, and social requirements, such as the demands of other stakeholders – from employee, to local communities, to society at large (cf. Gunningham 2009: 226).

conceptualisation of discourse as socially constitutive, as well as socially reflexive (Fairclough/Wodak 1997: 258), and views the rise of CSR as an example of discourse-led social change (Chouliaraki/Fairclough 1999) entailing – amongst other things – the recontextualisation and reframing of existing discourses and practices. More specifically, the rise of CSR to discursive hegemony involves the recasting of business discourse in socially responsible terms, with a rhetorically-driven blurring of boundaries between the discourse of profit on the one hand, and, on the other, that of socio-environmental responsibility, which can thus become the primary source of self-representation as well as self-legitimisation.

In this respect, Monsanto is a particularly interesting company to study. A world-known US-based biotech multinational, over the last few years Monsanto has invested heavily in communication campaigns that focus on sustainability-related themes. The company operates in a business sector which is especially subject to charges of environmental unfriendliness, since it involves the production, amongst others, of pesticides, bovine growth hormone, and genetically modified seeds, all products whose environmental safety is controversial to say the least. In addition, Monsanto has often been the object of intensive media attention because of its aggressive marketing and patent policies, and has been embroiled in highly mediatised lawsuits which have threatened to tarnish its reputation. It is therefore crucial for the company to persuade the public at large of the environmental safety and social beneficence of its core business, so that the consensus around biotechnology can be increased.

The communication campaigns mentioned above are part of this far-reaching consensus-building operation, which is characterised by a consistent deployment of legitimisation strategies based on CSR-related values. These strategies are ubiquitous in Monsanto's communication, and are especially evident in the company's Pledge reports, which are documents issued yearly to report on the community- and environment-related programmes Monsanto runs around the world.[3] By reason of their continuity of publication, the

3 The name 'Pledge' report is a reference to Monsanto's Pledge, a code of conduct adopted by Monsanto in 2000 which highlights the company's

reports represent a privileged site for the diachronic investigation of the exploitation of CSR in corporate communication for consensus building purposes. Their analysis may be expected to provide insights into the evolution of CSR discourse in an especially controversial sector, where assuaging the general public's scepticism and winning support for the sector's core activities is particularly important.

Based on the above, the specific aim of the investigation can be summed up in two separate but related research questions. The first one is a preliminary question, and concerns the genre status of the Pledge report – its communicative purpose and its function within the broader scenario of corporate reporting genres; the second concerns its evolution over time. The two questions can be formulated as follows:

1. can Monsanto's Pledge report be considered as a representative sample of the genre of the sustainability report? What are the distinguishing features of the Pledge report – if any – *vis-à-vis* those of the sustainability report? And
2. what are the textual, discursive and rhetorical strategies deployed to fulfil the genre's communicative purpose, and how do they change over the period considered?

2. Corpus and method

With respect to the materials on which the investigation has been conducted, it has to be pointed out that Monsanto's corporate website only features the latest issue of the Pledge report, the first one of which published in 2000; however, it was possible to retrieve previous issues from the Web, where they had obviously remained posted even though links are no longer provided in the dedicated section in Monsanto's website. Table 1 below shows details of the corpus:

---

commitment to environmentally and socially responsible behaviour. Monsanto's Pledge can be found at <http://www.monsanto.com/whoweare/pages/monsanto-pledge.aspx> (last accessed 10/02/2012).

	1999-2000	*2000-2001*	*2001-2002*	*2003-2004*	*2004-2005*	*2006-2007*	*2007-2008*	*2008-2009*	*Overall*
Tokens	25,970	22,907	14,855	17,338	15,085	12,880	12,307	17,999	*140,341*
Standard. TTR	39.92	40.10	43.89	42.08	42.73	42.61	41.62	41.85	*41.59*

Table 1. Details of the Monsanto Pledge report corpus.

The corpus includes all reports issued between 2000 and 2009, with the exception of the report for the period 2002-2003, which could not be located. The documents are fairly variable in length, with early ones featuring a much larger number of words compared to later ones. Their lexical density increases slightly with the reduction in their size, even though the changes are not particularly significant, and there appears to be overall consistency throughout the period.

For what concerns the methodological approach adopted, the investigation is located within an overarching discourse-analytical perspective and relies on a multi-method toolkit comprising genre analysis (Swales 1990, Bhatia 1993), text linguistics (Werlich 1983), argumentation theory (van Eemeren/Grootendorst 1992) – with special regard to strategic manoeuvering (van Eemeren/Houtlosser 1999, 2002, 2006) – and, wherever expedient for the analysis, corpus linguistics.

In particular, the genre-analytical framework was deemed to be especially useful for the study of the generic integrity of the Pledge reports *vis-à-vis* standard versions of the CSR report. A hypothesis that the research attempts to test, in fact, is whether sectors subject to particularly heated controversies may display, in their CSR reports (of which the Pledge report is an instantiation), features which set them apart from the standard versions of the genre.

Text linguistics and argumentation theory have been relied on to investigate the textual resources deployed in the reports. Text linguistics has made it possible to identify the various text types (and their interplay) featured in the reports, and link them to the persuasion strategies enacted, for the analysis of which recourse has been had to elements of argumentation theory.

Finally, the investigation has relied on automated text interrogation routines (Wordsmith Tools 4.0) to retrieve significant lexical items and isolate recurrent patterns within and across reports, with a view to identifying their diachronic evolution.

3. Pledge reports and generic integrity

The first question to be answered concerns the genre status of the Pledge report. As mentioned above, the Pledge report appears at first sight as an instantiation of the better known CSR report, an emerging but increasingly widespread document used by companies to report about their social and environmental performance. Such interpretation is due to two main factors. First, companies typically publish two types of report – the annual report, which deals with the financial performance of the company, and the CSR report, which focuses on social and environmental performance; by default, therefore, a report which is *not* the annual report and which deals with social and environmental themes is likely to be interpreted as some kind of CSR report. Secondly, the name given to the report – the Pledge report – while not standard, evokes the value-based approach which is a defining feature of CSR discourse, thereby reinforcing the impression that the document does indeed belong to the genre of the CSR report. But is it really so? More specifically, is the function of the Pledge report the same as that of the 'standard' CSR report? Further, has the Pledge report, as a genre, evolved over the years?

3.1. Pledge reports and communicative purpose

Insofar as communicative purpose is a key genre determinant, identifying the communicative purpose of the Pledge report *vis-à-vis* the CSR report can help clarify the issue of its genre status. The overall communicative purpose of the CSR report is to provide data about the company's social and environmental performance, with a view to enhancing its reputation and providing it with licence to operate. Compared to this, the communicative purpose of the Pledge report appears to be much broader. Its main characteristic appears to be an all-encompassing preoccupation with the promotion of biotechnology *per se* as a means to foster sustainability at a global level, and only on a lesser scale with the provision of information about the company's eco-efficiency and social commitment. Pledge reports argue again and again for the intrinsic social and environmental benefits of biotech crops, which are of course best delivered through the company's active engagement with farmer communities (especially in third-world countries), but have an intrinsic value *per se*. Thus, the description of the company's commitment to social responsibility in its own operations is subsumed under a much broader legitimising objective, which is the production of genetically modified seeds. The two objectives are acknowledged to be synergic, as a quote from the 2001-2002 report highlights:

(1) We know that our success as a business and society's acceptance of biotechnology will hinge in part on how well we take account of and contribute to the advancement of broader public interests.

Still, should support for biotechnology as such fail, it would be nearly impossible for the company to obtain the licence to operate which it needs to be a successful business. The Pledge report becomes therefore a powerful lobbying instrument for the defence of the company's core business, rather than simply for the company as such.

The broader scope of the Pledge report as a genre is reflected in its rhetorical organisation. Compared to the standard CSR report, which in the model put forth in Catenaccio (2011a) comprises four moves (1. Signalling understanding of CSR as part of business operations; 2. Signalling commitment towards environmental and social issues; 3. Signalling good business practices; 4. Establishing credentials), the Pledge report features a structure which comprises, besides these moves, an extensive section highlighting the social and environmental benefits of Monsanto's products and of its biotech dissemination projects, all of them framed in CSR terms. This section may be interpreted as being subsumed under the description of the company's business practices (Move 3), but its size and salience in the report suggest that its classification as a self-standing move may be more appropriate. In conventional CSR reports Move 3 is devoted to an in-depth account of the CSR management process, with extensive data on eco-efficiency, human resources management and safety issues, to which details of further CSR-related initiatives are then added. In the Pledge reports, by contrast, the management process is only briefly sketched, while the lion's share is taken by product and project descriptions, which are invariably made to precede the other conventional data. Thus, the move structure of the Pledge report may be summed as follows:

1. Signalling understanding of CSR as part of business operations
2. Signalling commitment towards environmental and social issues
3. Arguing in favour of biotechnology (by giving evidence of its environmental and social benefits)
4. Signalling good business practices
5. Establishing credentials

Despite being formally akin – by and large – to the CSR reports issued by the majority of companies, therefore, Monsanto's Pledge reports display additional features which have the effect of shifting their communicative purpose from the description of the company's social and environmental performance to the promotion of biotechnology. The broadening scope of the document is acknowledged in the first of the Pledge reports, whose *incipit* is reported below (emphasis added):

(2) Monsanto is an agricultural solutions company, supplying farmers worldwide with several herbicide products including Roundup.
We are also a supplier of seeds for a variety of crops, some of which are bioengineered to be insect-protected and/or tolerant to Roundup herbicide.
As you read through this report, you'll find information about Monsanto's performance in the areas of environment, safety and health (ES&H) and sustainable development. Monsanto has issued environmental reports since 1991. The reports have included data about the company's impact on the environment through emissions along with data about the safety and health of our employees, releases and transfers of wastes, fines, penalties and other information. *This report* also contains that traditional information but this year *expands beyond those areas to report on eco-efficiency, look at the impact of the use of our products and review some of Monsanto's programs and*

> *activities which have an impact on social and economic development in selected areas of the world.* (Monsanto Pledge report 1999-2000)

Over the years, this 'additional information' has become more and more prominent, coming to occupy the best part of the reports and making the above mentioned shift in communicative purpose ever more evident. Because of their subtle exploitation of the generic resources for the genre of the CSR report to pursue a different communicative purpose, the Pledge reports can be seen as an example of genre bending (Bhatia 2004: 86-88). As Bhatia (1993: 13) points out, the constraints which genres impose on texts can be "exploited by the expert members of the discourse community to achieve private intentions within the framework of socially recognised purposes". The Pledge reports exploit these constraints in an extremely sophisticated way, effectively turning a genre of corporate reporting into one of cultural propaganda.

3.1.1. From reporting to explicit arguing

The exploitation of the features of an established genre to achieve a different communicative purpose from the one conventionally associated with it seems to increase over the years considered, with a gradual growth in the space and salience given to texts devoted to extolling the benefits of biotechnology accompanied by a reduction in the commentary on eco-efficiency. Sections explicitly devoted to argumentative discussions on biotech topics are evident in all the reports except for the first one, and are organised as a two- or multi-part debate in which a particularly contentious aspect of biotechnology is addressed from different perspectives. Typically, the discussion is introduced by a passage which mentions the main points of the debate (generally in a rather biased manner, despite its factual tone), followed by the positions of both detractors and supporters of biotechnology. In the 2001-2002 report the dilemma tackled is that of food labelling for food containing genetically modified ingredients. The issue is introduced by the passage reported in excerpt (3) below, and is followed by three separate texts presenting each a different position: first, the opinion of many consumer groups ('Consumer's perspective') (ex. 4) is given, followed by the 'Industry's perspective' (ex. 5) and finally by 'Monsanto's approach' (ex. 6), which is given the final word. Each position is presented independently and is clearly attributed (i.e., the 'voices' of the players in the debate are explicitly introduced), and no concluding remarks are provided, thereby ostensibly leaving it for the readers to draw their own conclusions as to what position is more convincing. Of course, despite its seemingly balanced structure, the section can hardly be said to be neutral. Rather, it is rhetorically constructed to lead the reader to favour Monsanto's position. This is evident from the very beginning, as shown below:

(3) *The dilemma*

Few issues are as contentious as the debate over whether or not to note biotech ingredients on food labels. Labeling proponents opine that consumers have a basic right-to-know which foods have biotech ingredients so that they can choose what they eat. Others believe that labels should be limited to relevant nutritional and safety information as the United States Food and Drug Administration now requires. Other information – such as how the food is grown or processed – should be available through other sources (such as toll-free information lines) and the subject of consumer education.

In the US, the Food and Drug Administration has determined that if biotech foods are comparable in ingredient composition, functional properties, nutritional profile and safety to conventional foods on the market, then there is no relevant information to place on a label. The same labeling requirements that apply to food in general apply to food derived from plants modified through biotechnology, i.e. the process of breeding the plant is not a basis for labeling.

Some countries, such as Japan, have instituted labelling laws for food with biotech ingredients. These regulations may require segregation of biotech and non-biotech crops throughout the grain handling and food processing systems, which adds extra expense that must eventually be passed along to the consumer. Europe's proposed regulations around traceability and labelling are even more costly and burdensome.

The passage opens with the presentation of the two conflicting positions upon which the whole debate hinges. The way in which they are introduced is characterised by an evaluative component, with the pro-labelling side being said to "*opine* that consumers have a [...] right to know", whereas their opponents "*believe* that labels should be limited to relevant nutritional and safety information" (emphases added). While 'opine' means "to say one's opinion", it is often used in formal contexts and frequently implies a negative evaluation of the person holding the opinion, who is regarded as pompous and pedantic (*Webster's New World College Dictionary*, 2010, *ad vocem)*, 'believe' is typically endowed with a positive connotation, indicating commitment to the truth value of the following proposition, and thereby shedding a positive light on the holder of the belief. The choice of these reporting verbs contributes from the very beginning to framing the debate in favour of the anti-labelling party, whose position is implicitly supported by being presented as aligned with the requirements of the Food and Drugs Administration. In the lines that follow, the passage prepares the ground for the promotion of the voluntary provision of information in dedicated spaces other than food labels (the position which is later revealed to be favoured by Monsanto). This position is presented as linked to the anti-labelling argument, but is embedded in a self-standing syntactical unit which – differently from the previous two – is not clearly attributed, but can instead be interpreted as being averred (Sinclair 1986, 1988), therefore gaining in epistemic strength. The following paragraph ("In the US … as a basis for labelling") is an expository passage which explains current labelling requirements, whereas the last paragraph introduces a new element in the debate – the issue of the added cost of labelling – which, while not pertaining strictly speaking to the topic under discussion, is likely to strike a particularly sensitive chord in the general reader, thereby contributing to steering him/her towards the anti-labelling position.

The different positions mentioned in the introduction are given a more extended treatment in the following page, where the following three texts appear:

(4) *Consumers Have a Right to Know*
By Edward Groth III, Ph.D.,
Senior Scientist, Consumers Union

When it comes to what we eat, consumers demand choices. Sustainable food production is just one factor that consumers-as-citizens care about. Among the others:

Right-to-Know – Consumers have a right to know what they are eating. Labeling that clearly identifies genetically engineered foods and ingredients

would both serve that right, and make any perceived risks associated with genetic modifications voluntary, and thereby far less worrisome.
Freedom – Consumers have a right to choose. Labeling is essential so consumers can exercise that right.
Trust – Consumers generally trust that the foods they see in the supermarket are safe and wholesome. When doubts are raised though, they tend not to trust industry or Food and Drug Administration assurances that everything is perfectly safe. Refusing to label foods as containing genetically engineered materials suggests that the producers are trying to hide something, and increases distrust.

(Excerpted from 'The Debate Over Food Biotechnology in the United States: Is a Societal Consensus Achievable?', *Science and Engineering Ethics*, Volume 7 pages 327-346, 2001).

(5) *Accurate and Informative Food Product Labeling*
By Grocery Manufacturers Association

Labels on all packaged food must provide consumers with truthful, clear and non-misleading information. For this reason, we support the current Food and Drug Administration (FDA) policy, which requires, among other things, that food be labeled if it poses a safety concern, as is the case for all foods. Safety issues can arise where food produced through biotechnology or other means is significantly different from its traditional counterpart.
Every food product on a grocery store shelf must meet the law's requirements for safety and labeling. Federal regulation has never required that food labels describe the plant development process by which food is produced. Such processes are not material information that must be labeled. Because the law is aimed at ensuring safe food and informative labels, biotechnology products should not be singled out for special regulatory treatment unless there is a significant difference in composition or, otherwise, a safety problem exists, or where material information is missing.

(6) Monsanto's approach
Supporting the Voluntary Labeling Guidelines

Monsanto believes that people who want to eat food that does not contain biotech ingredients should have that option. In the United States, organic products are produced using practices that should assure that there is no intentional presence of biotechnology products. Organic foods therefore provide a ready source of foods that do not contain biotechnology ingredients.
Monsanto remains in active dialogue with many people and groups about this issue, and is committed to making product information available and

> accessible. For people who want to know more about Monsanto's biotech crops, we have made hundreds of studies on their safety and environmental impact available on our website. Some food companies also provide more information about their food products on their own websites. We want to do our part to provide consumers with relevant information that they need to make healthy food choices, and to ensure food systems around the world operate as efficiently and effectively as possible to bring food to consumers in a cost effective way. Those priorities are built into the current system and must be weighed before any changes are made.

Of these three passages, the one presenting Monsanto's approach is especially interesting because it clearly shows the company's strategic use of argumentative devices. The passage opens with a concession to the pro-labelling proponents ("Monsanto believes that people who want to eat food that does not contain biotech ingredients should have that option"), which is exploited to establish a shared starting point upon which to build its argument. This use of concession is very useful in argumentation, since – as Perelman and Olbrechts-Tyteca (1969: 488) point out –

> it is a good thing never to refute [one's opponent's] arguments in such a way that he seems a poor advocate. [This can be done through concession, which] expresses the fact that one gives a favourable receipt of one's opponent's real or presumed arguments. By restricting his claims, by giving up certain theses or arguments, a speaker can strengthen his position and make it easier to defend, while at the same time he exhibits his sense of fair play and his objectivity.

After this initial concession, Monsanto goes on to argue that the "freedom of choice" the pro-labeling party insists on is not threatened, as consumers can opt for organic products, which cannot by definition contain genetically engineered ingredients. By putting forth this argument, Monsanto shifts the burden of proof for the complete labeling of food from the mainstream food industry to organic food manufacturers, adding in the process an element of doubt about the truthfulness of the latter's claims to naturalness ("organic products are produced using practices that *should* assure that there is no intentional presence of biotechnology products", emphasis added). In the remaining part of the passage Monsanto declares its willingness to provide information about biotech products, and closes by insisting on the cost-effectiveness of current food manufacturing methods (which include the use of genetically modified ingredients), thereby evoking the risk of heightened costs for food and suggesting – without stating it openly – that compulsory labeling would be against the consumers' own interest. The exploitation of the economic motive represents, I believe, a clear exemplification of what van Eemeren and Houtlosser (1999) call strategic maneuvering through the exploitation of the *topical potential* of a discussion, i.e. through the selection of materials from those available according to what the speaker believes is likely to best advance his/her interests. In this case, appealing to the risk of heightened costs for foodstuff represents a clear attempt to present common readers with an argument to which they can easily relate, thereby increasing the persuasive power of Monsanto's position.

This discussion, however limited, shows the extent to which explicit argumentation features in the Pledge reports. This is a peculiar feature of the Monsanto reports, and sets them apart from more conventional examples of the genre. Moreover, the themes selected for explicit argumentative discussion are eminently topical, and testify to the evolution of the discourse around genetic engineering throughout the period discussed. By looking at them it is possible to trace the key concerns that the biotech industry has had to face over the years, as well as the argumentative strategies most commonly deployed to assuage the fears of biotechnology detractors.

3.2. Recurring arguments, evolving discourses

As highlighted above, the communicative aim of the Pledge reports is therefore, on the one hand, to build the case for biotechnology as such, and, on the other – through the demonstration of the legitimacy and beneficiency of its core business – to discursively construct Monsanto's 'licence to operate'. In Section 3.1.1 I have described the way in which explicit argumentation features in dedicated sections in the reports. In this section I intend to further explore the use of argumentative strategies in the Pledge reports, extending the analysis to those parts of the texts which are not explicitly framed as argumentative discussions.

The first aspect I shall consider concerns the presence in the documents of a recurrent argument scheme geared at the promotion of biotechnology which is recursively deployed throughout the period considered; I will contend that it represents a constant element in Monsanto's communication strategy, guaranteeing argumentative consistency across thematic variation.

The second issue I shall discuss centres on topic development. By examining the reports diachronically it is possible to trace the evolution of selected topics over the years. Such evolution, I shall argue, is grounded in a conceptualisation of the reports as a means to obtain readers' consensus on biotechnology by 'educating' them – a communicative purpose which endows the reports with an intrinsically argumentative function.

3.2.1. Arguing biotechnology: a pragmatic approach

If we accept that the overarching communicative purpose of the Pledge reports is to gain consensus for biotechnology, we can also see how they are fundamentally geared at persuasion. According to Perelman (1982: 5), persuasion can be seen as part of a more general notion of argumentation, which "covers the whole range of discourse that aims at persuasion and conviction, whatever the audience addressed and whatever the subject matter" (cf. also Virtanen/Halmari 2005: 5). From this perspective the Pledge reports can be seen as macro-arguments in defence of a position that can be formulated as follows:

> Biotechnology should be used in agriculture

A typical example of the arguments put forth to support such position is provided in example (7) below:

(7) Monsanto's approach to business and the environment is based on the understanding that the world is going to need a lot more food in the future and that the food systems of the world will have to change to meet that need. We believe that one part of meeting that need is the use of biotechnology to redesign the way food and fiber is produced. This is part of our vision for the future. (Letter from the CEO 1999-2000)

The argument structure of the passage above can be schematically represented as follows:

> Biotechnology should be used in agriculture
> *because* using biotechnology in agriculture will make it possible to meet the world's growing food needs
> [*and* meeting the world's growing food needs is desirable],

where the last proposition is an unexpressed premise, as highlighted by it being encased in square brackets. This is an example of a well-known argument scheme which in pragmadialectic theory is called 'pragmatic argumentation' (van Eemeren/Houtlosser/Snoeck Henkemans 2007: 183), and which can be represented in more abstract terms as shown below:

> X should be done
> *because* X leads to Y
> *and* Y is desirable.

In the case at hand its persuasive power is enhanced by reliance on an especially strong object of agreement – the desirability of producing enough food and fibre to meet the world's growing needs. As Perelman and Olbrechts-Tyteca (1969: 65) point out, selecting a strong object of agreement is crucial in argumentation, as

> the unfolding as well as the starting point of the argumentation presuppose indeed the agreement of the audience. [...] When a speaker selects and puts forward the premises that are to serve as the foundation for his argument, he relies on his hearers' adherence to the proposition from which he will start.

Monsanto's arguments in support of biotechnology typically rely on premises of this kind – the desirability of the eradication of hunger, or of agricultural production methods which reduce soil erosion, or which enable small farmers in third world countries to produce enough food not only to survive, but also to improve their social status – all of which are likely to be broadly shared by the audience. In fact, so universally accepted are these premises that they typically do not even need to be expressed in the argumentation, but are implicitly relied on it its unfolding. Another example of argumentation based on the pragmatic scheme illustrated above is illustrated in excerpt (8), where the unexpressed premise is the desirability of increased environmental stewardship coupled with reduced costs (an object of agreement likely to strike a particularly sensitive chord with the farmer community):

(8) The Red River Valley in Manitoba, Canada, is a fertile flood plain that presents daunting challenges for farming: adverse weather and frequent flooding. These challenges and the struggle to remain profitable are daily considerations for ART ENNS, who has farmed in the valley for 30 years with his brother and their two families. They plant 3,000 acres of canola and soybeans each year.

The Enns partners share a love of farming and the land. Farming "gives you the independence of being stewards of the land," says Enns. "Not only taking out of the land, but providing so the next generations will have a future in farming."

Enns has used herbicide-resistant biotech crops to help deal with the challenges the Red River Valley poses for eight years. "Because of excessive rains and flooding, our crops were under pressure from all kinds of different weeds," he says. "Controlling them with conventional methods was very difficult....We were actually at times mixing two, three chemicals together and having to go back three, four times to try to get control of the weeds."

With biotech crops, Enns says he has reduced chemical spraying to just one pass, which results in more profitability and better environmental stewardship. "Yes, it does save us some money, but it also means less application of chemicals, less chance of any contamination or runoff that may pollute," he says.

Using the scheme described above, the argumentation presented in this passage can be represented as follows:

Biotechnology should be used in agriculture
because using biotechnology in agriculture leads to increased profitability and better environmental stewardship
[*and* increased profitability and better environmental stewardship are desirable]

As in the case of passage (7), the desirability of the effects brought about by the use of biotechnology can hardly be disputed. Moreover, by couching such arguments in typical 'success stories', Monsanto greatly enhances what Perelman and Olbrechs-Tyteca call their *presence* (1969: 142), which relies on "the displaying of certain elements on which the speaker wishes to center attention in order that they may occupy the foreground of the hearer's consciousness".

By selecting arguments which rest on broadly shared assumptions, the company is operating a very effective strategic choice: it poses biotechnology as the answer to mankind's most pressing societal and environmental problems, thereby framing it as essentially benevolent and effectively obliterating its business component.

3.2.2. The argumentative use of exposition

A further way in which Monsanto attempts to build consensus for biotechnology is through the 'education' of the general public, whose diffidence is seen by the company as the direct consequence of ignorance, or mis-information. On several occasions both in the reports and in Monsanto's website communication is the claim made that better information is needed in order to persuade the public at large of the benefits of biotechnology. The passage below, illustrating the communicative aim of the *Transparency* section of the *Monsanto* Pledge website, is a case in point:

(9) The 'Transparency' section of the new Monsanto Pledge Web site also features comprehensive lists of published scientific literature on benefits associated with agricultural biotechnology. [...]
"We're committed to making information available and more accessible to people who are interested in the technology but may not be able to easily access all of the published information," said Roy Fuchs, director of regulatory science for Monsanto and leader of the company's Transparency working team. "We believe making this information more accessible has significant value for the scientific community and the public because *access to comprehensive and scientifically sound information is essential for advancing an informed discussion about this technology*," he said. (2000-2001 Pledge Report)

The fact that the research published on the website only reports on the *benefits* of biotechnology does not seem to trouble the writer in the least. In the same Pledge report another claim is made (attributed to an independent academic, which enhances its authoritativeness and at the same time downplays Monsanto's vested interest) highlighting the link between consumer education and consensus:

(10) Transparency serves several purposes. In addition to the obvious one of keeping companies and government agencies accountable to the consumer, *transparency helps build consumer confidence and eases the acceptance and adoption of new technologies*. (Emphasis added)

In the following issue of the Pledge report this connection is made even more explicitly (and one more time it is attributed to an external, independent voice):

(11) "[...] The challenge for Monsanto is to use what it learns not as ammunition to strengthen its capacity to fight any non-biotech believers, but to develop a sound business strategy that creates value for its own shareholders in harmony with societal needs and concerns. [...]
Participating in the work of the council is a risk for us and for the company. To me, the Monsanto Pledge and the council itself represent a courageous initiative toward earning public trust and transforming a corporate culture. [...] Local Monsanto appears to have internalized the spirit and behavior embodied in the Pledge. However, stakeholders beyond the company have yet to appreciate the significance and sincerity of the commitment. *Changing behavior and changing perception will take a little time, but there must be no turning back,* for we live in the realities of a redefined bottom-line." (Emphasis added)

These passages clarify the overall aim of Monsanto's communication strategy within and beyond the reports, emphasising the perceived link between information and persuasion, and suggesting that ostensibly informative (and more specifically expository; cf. Werlich 1983) passages serve an overarching persuasive intent.

3.2.3. A diachronic view of topic-internal variation

In light of the above, therefore, it may be expected that analysing the way in which information on selected topics is provided report after report will shed light on the long-term strategy pursued by Monsanto. A preliminary reading of the reports suggested that they featured several recurring topics whose handling underwent subtle variations over the period considered. In order to identify significant topics and examine the implications of their variation, two complementary methods were used: a keyword study (which identified lexical items which occurred with a particularly high frequency in each of the reports compared to all the others) and close reading, which made it possible to pinpoint differences in the treatment of topics which appeared in all the reports, but with varying (typically decreasing) frequency over the period.

A complete keyword analysis clearly goes beyond the scope of this chapter, as it would require extensive discussion. Suffice it to say that amongst the keywords identified are the initialism *BT* (2004) (which stands for Bacillus Thuringiensis, a common soil bacterium; Monsanto markets pest-resistant seeds containing a gene from this bacterium) and words such as *sequence* (2000-2001), *ethanol* and *biodiesel* (2006-2007), and *yield* (2008-2009). Of course the keyword status of these terms is determined by the salience that a topic is granted in a given year, which may be due to contingent factors (for instance, for what concerns the word *sequence*, the reason for its salience is to be found in the reference to the completion, in 2000, of the rice genome sequence, which prompted an extensive discussion of the importance of sequencing the genomes of common crops). However, in the cases which were amenable to longitudinal consideration, comparable patterns of diachronic variation were identified. To best illustrate such pattern, it was decided to focus on a single lexical item (*tillage*) which has keyword status in the first Pledge report and then gradually decreases in frequency over the following years, as shown in Table 2 below.

Report	*Ranking*	*%*
1999-2000	28	0.36
2000-2001	36	0.30
2001-2002	131	0.08
2003-2004	155	0.09
2004-2005	136	0.10
2006-2007	378	0.04
2007-2008	295	0.04
2008-2009	2,847	/

Table 2. Ranking and relative frequency of the word *tillage* in the Pledge reports 2000-2009.

The word *tillage* first occurs in an extensive passage dedicated to the topic in Monsanto's 1999-2000 Pledge report, which is reported in example (12) below:

(12) [C]onservation *tillage* [*is*] a farming practice that utilizes little if any *tillage*, leaving the soil relatively undisturbed. It's not a new concept and some farmers have used it for years as a way of reducing erosion, protecting the

> water quality of streams and lakes and providing habitat and food for birds and other wildlife. Now however, farmers are learning that conservation *tillage* can significantly mitigate global climate change. The concept is simple: through the photosynthesis process, crops pull carbon dioxide from the atmosphere and store it in their stalks and leaves. When their residue is left on the fields to decompose, the carbon they pulled from the atmosphere is deposited in the soil. Unfortunately, traditional intensive-*tillage* farming practices release virtually all of the carbon back into the atmosphere. *Tillage* stirs up the soil and exposes the stored carbon to oxygen, which turns it into carbon dioxide and releases it back into the atmosphere. However, if the soil is farmed with conservation *tillage* techniques, a significant portion of the carbon sequestered by the crop is returned to the soil profile building up the soil organic matter.
>
> Monsanto's agricultural products are well suited to support and encourage conservation *tillage*. Roundup herbicide is an environmentally-friendly tool that farmers can use to control weeds without resorting to the plow or cultivator. Roundup has been instrumental in facilitating conservation *tillage* around the world.
>
> The more recent addition of Roundup Ready crops has further encouraged and facilitated the adoption of conservation *tillage*. These crops are resistant to Roundup herbicide which means that even after the crops are up and growing, Roundup herbicide can still be used to control weeds without *tillage*. Monsanto believes that it can help mitigate climate change through its products, especially by encouraging the adoption of conservation *tillage*. As a company, Monsanto is also committed to support and help the agricultural community as a whole adopt practices that sequester carbon and develop technologies that are climate friendly. (1999-2000)

The sequence opens as an expository (Werlich 1983) text type, with a typical definition, expressed by means of a simple phenomenon-identifying sentence ("conservation tillage is a farming practice that utilizes little if any tillage, leaving the soil relatively undisturbed"). The definition, however, is embedded in a heavily evaluative structure (Hunston/Thompson 2000). Conservation tillage is said to be a "simple" concept, and emphasis is placed on the fact that it is "not new". These remarks are clearly designed to address the fears of sceptics who are likely to look at biotechnology as something which represents a break with traditional farming practices, and involves extremely complicated procedures. Establishing a link with time-honoured, if not especially widespread, techniques aimed at sustainability and environmental preservation therefore helps frame the use of biotech crops as innocuous. In the following lines the evaluative component is further enhanced through the use of the adverb of stance *unfortunately*, which introduces a sentence highlighting the undesirable environmental effects of conventional agriculture (the one detractors of biotech would like to defend). Having thus painted a fairly bleak picture of the current state of agriculture, the text goes back to the theme of conservation tillage, insisting on its potential to foster sustainability. The mention of such potential is embedded in a concessive structure which introduces an element of counter-expectation; in other words, if the fist part of the passage gives a pessimistic view of the overall impact of agriculture on climate change, the second one indicates a little known but potentially highly effective path leading to a brighter future. It is only at this point, having established a need for conservation tillage as a way to reduce carbon emissions in agriculture, that the role of Monsanto is introduced, and the ultimately argumentative aim of the passage becomes apparent. Once again, the underlying argument structure is the one illustrated in Section 3.2.1, although the details of the argument itself change:

> Biotechnology should be used in agriculture:
> *because* using biotechnology in agriculture encourages conservation tillage
> *and* conservation tillage is desirable

Differently from the examples provided in Section 3.2.1, though, the premise that conservation tillage is desirable cannot be considered broadly shared by the audience. This is the reason why an extensive explanation of what conservation tillage is and why it is desirable is in place. This explanation has a didactic purpose – it explains what is meant by conservation tillage – but at the same time is designed to establish a point that may be used as an object of agreement in later argumentation. Thus, the explanation provided in the 1999-2000 report is never re-proposed. In later reports, the beneficial effects of conservation tillage are increasingly treated as a given, and are used in argumentation without further elaboration.

(13) Monsanto provides several products that facilitate conservation *tillage*, including Roundup agricultural herbicide, Residue Proven seeds and Roundup Ready crops. These products enable farmers to avoid all *tillage* if they want to. Monsanto Company has been encouraging conservation *tillage* for many years

because of the economic and environmental benefits it provides. In conservation *tillage* systems, growers minimize soil disturbance. (2000-2001)

(14) Monsanto's products can help in important ways. Our Roundup agricultural herbicides are useful for conservation *tillage*. They let farmers control weeds without plowing or hoeing, thereby saving backbreaking labor and conserving soil and water. (2005-2006)

(15) We have already developed plants that result in reduced pesticide use, better weed control, and increased conservation *tillage*, while steadily increasing yields. (2008-2009)

In all these excerpts knowledge of the reasons why conservation tillage is desirable is taken for granted, or only superficially hinted at (see examples 13 and 14). The 1999-2000 report is the only one in which a full explanation is given: in later texts both the expository sequences and the explicit declaration of why conservation tillage should be considered desirable disappear, replaced by textual usages that rely on implicit knowledge of both conservation tillage as such, and of its importance for environmental protection. This suggests that successive issues of the report rely on a body of knowledge established in the previous ones for the success of their communication strategy, assuming, moreover, that this knowledge has been internalised by the reader, whose alignment with Monsanto's position in respect with the desirability of conservation tillage is taken for granted.

4. Conclusions

The diachronic study of Monsanto's Pledge reports discussed in this chapter has shown that even while the genre as such has not undergone dramatic changes over the years, it seems nonetheless possible to detect in subsequent issues subtle signs of a progressive increase in the argumentative component which is its distinguishing feature, and which sets it apart from the genre of the CSR report to which it is akin.

Compared to the CSR report, the Pledge report has a broader communicative purpose, which is to gain consensus for biotechnology. In this sense, it can be seen as an instance of genre bending in the pursuit of private – or rather industry-specific – intentions. This broader communicative purpose is not only reflected in the dedicated argumentative section referred to above (which explicitly tackles controversial issues), but also has an impact on the overall discursive organisation of the reports, which are largely devoted to demonstrating the benefits of biotechnology.

In order to demonstrate these benefits, the reports make recourse to a standard argumentation scheme – called pragmatic argumentation in pragmadialectics – which is deployed recursively across them. The (often unexpressed) standpoint that the texts aim to support is that biotechnology should be used in agriculture. The arguments put forth in defence of this standpoint highlight the beneficial effects of specific uses of biotechnology on a variety of momentous issues – from the eradication of hunger, to environmental protection, to poverty reduction. The consistent reliance on the same argument scheme in support of an equally unchanging standpoint represents an element of continuity which is matched by a constant variation in the supporting arguments put forth, which, in turn, testifies to the progress of biotechnology in the provision of ever better solutions. Thus, by looking at the progression of the supporting arguments, it is possible to trace the evolution of biotechnology over the period considered. Moreover, the fact that in subsequent issues of the Pledge reports attention is paid to providing ever new arguments suggests that a long-term strategy is in place which focuses on evolution and deliberately refrains from insisting on points which have been discussed in previous issues, and which can therefore be taken for granted.

Evidence of this diachronically oriented strategy can also be found in the way in which specific topics are developed report after report. Whenever a new topic or concept is introduced, a full explanation is given for the benefit of the layperson who may be unaware of the technicalities of biotechnology. These explanations, however, only occur upon the first mention of the topic. In subsequent issues of the Pledge report knowledge of the topic is taken for granted, and often implicitly relied on for the purposes of argumentation.

The strategies described above suggest that through its Pledge reports Monsanto pursues a strategy of persuasion through the gradual establishment of consensus on a fairly small number of selected issues. Each year a single new topic is tackled, while previously discussed issues are never raised again. In this way they can be considered – at least in the textual reality of the Pledge reports – as no longer contentious, and assumed as being part of an ever broader shared basis of consensus which does not need to be real in the outside world, but is treated as real in the discursive environment of the reports, thereby conferring upon biotechnology a sheen of acceptability which may be in part fictitious, but which is convincingly argued and may eventually succeed in changing the public opinion on the topic.

CINZIA GIGLIONI

Apologetic Discourse in Financial Reporting: CEO and Chairman Statements. A Case Study

1. Aim and method

This chapter focuses on the deployment of apologetic discourse as a generic resource in Annual Company Reports, looking at its variation as a function of contextual factors connected with corporate performance. The study of apologetic discourse has a long track record in the history of rhetoric, spanning from the writings of Plato and Aristotle to contemporary publications such as Ware/Linkugel (1973) and Benoit (1995) and, more recently, Koesten/Rowland (2004). While occurring in a variety of discourses, apologetic strategies have been most fruitfully analysed in the context of political discourse (cf. Harrell/Ware/Linkugel 1975; Harris/Grainger/Mullany 2006). In recent years, however, a growing interest for image and reputation management in the corporate domain, coupled with an increasing awareness of its underlying rhetorical nature, has made the study of apologetic discourse particularly interesting for business communication scholars, as testified by the success of works such as Benoit (1995) and Hearit (2006).

Apology is of course typically deployed in crisis situations, when companies and individuals have to respond to allegations of wrongdoing. Fortunately, full-blown emergencies are not common events. In the business environment, however, where image and reputation are a crucial component of corporate success, it may be expected that constant attention be paid to every sign of potential trouble, however minimal, which may be perceived by the company's shareholders as an indication of failure on the part of the company itself to meet standards or expectations. In particular, companies can be expected to engage extensively in apologetic discourse when they are underperforming, and less frequently when they are performing well. In either case, the apologetic strategies deployed may vary, depending on the circumstances and on the type of reputation threat perceived. Investigating companies' use of apologetic discourse in a diachronic perspective – especially one involving a crisis vs non-crisis situation – may therefore shed some light on the incidence of apologetic discourse in corporate discourse in general, and in particular on its relative frequency in corporate reports in good and bad times.

Based on the above, the aim of this study[1] is to detect the use of apologetic discourse in CEO's and Chairman's Statements in annual company reports (ACRs), analyzing its use over time and identifying the role it plays, if any, in genre variation. For this purpose it compares ACRs issued respectively in 2000, the year before corporate accounting scandals and September 11, under stable market conditions, and in 2009 (at a time of crisis in the markets worldwide), looking at differences and changes. This period of time seems to be particularly suitable to study variation in this text genre as it has seen a generalized and marked deterioration in economic outlook at world level.

Annual reports are composite documents comprising separate and often self-contained sections. The distribution, structure and nomenclature of such sections may sometimes vary, but this does not prevent genre (and sub-genre) recognition (Garzone 2005a: 80) since active discourse community members are perfectly capable of recognizing the rhetorical actions performed by texts to which they are recurrently exposed (Swales 1990). Even the increasing 'promotionalization' of the annual report underlined by Bhatia (1997b, 2004) doesn't seem to really threaten its generic integrity. A diachronic perspective could therefore shed some light on this point since apologetic discourse, even if not strictly promotional, is generally aimed at company reputation defence. The study compares parts of the annual reports that can be identified as having a narrative function ('narratives' being the label used by consulting firms to refer to the verbal text of

1 Some of the preliminary results of this study have been presented in *IR Top*, a review for professionals in the field of Investor Relations (cf. Giglioni 2010).

company reports) from 2009 with their equivalents from 2000, thus taking a diachronic, crisis vs non-crisis perspective combined with a more general perspective on genre change and contextual factors.

1.1. Business communication and apologetic discourse

As anticipated above, numerous studies of apologetic discourse in general have been published over the past 35 years mainly by sociologists, psychologists and rhetoricians but scholars have begun to extend the categories of apologetic analysis to specific domains only in the last decade or so. Although some studies have dealt with corporate apology in a context where corporate practices have come under greater public scrutiny and companies function as apologetic rhetors, as pointed out by Jasinski (2001: 20-23), literature devoted to corporate apology has tended to concentrate on crisis communication in general (Benoit 1995; Hearit 2006) rather than on communication during economic and financial crises. In this study 'apology' will be taken into consideration in its very broad sense of justification of one's actions/position insofar as it is expected that such actions/position may be the object of potential criticism on the part of an imagined opponent. Thus, 'apology' involves justification of strategic choices; or defence of (poor?) results, and boosting of company's image.

The foundations for the present study are to be found in the findings of Ware and Linkugel's research published in 1973 (Ware/ Linkugel 1973: 273-282), which identified four tactics or common strategies in apologetic discourse: denial, bolstering, differentiation, transcendence. Other scholars have proposed alternative conceptualizations of the types of strategies, William Benoit (1995, 2000) among them. He identified five major strategies that overlap slightly with Ware and Linkugel's tactics: he included 'denial' as well, but the other strategies he recognized (such as 'mortification') are much more in line with image repair theory[2] than with the broader meaning of

2 Benoit (2000: 40) remarks: "I now tend to prefer image 'repair' to image 'restoration' because 'restoration' might imply that one's image has been restored to its prior state".

apology which has been adopted in this research. From this perspective, apology is not intended as a particular discursive response to a specific rhetorical attack, but rather as a recurrent type of discourse in a recurrent text, the annual company report.

Strategy definitions as used in this study, adapted from Ware and Linkugel (1973), are illustrated here below.

- Transcendence	is the strategy that tries to reframe the situation in a broader context.
- Denial	Ware & Linkugel distinguished among various forms of denial (denial of substance, of intent, of extent, of consequences) and forms of indirect denial: those are the forms of denial considered in the present study. Indirect denial is the response to charges in such a way that the charges are never explicitly acknowledged.
- Differentiation	is the strategy that seeks to create necessary distinctions that somehow redefine the questionable situation.
- Bolstering	is the process of enhancing the image of an institution (companies in our study) by linking it to abstract values.

The analysis of the deployment of these strategies seems to be especially useful to shed light on the mechanisms involved in genre change in the time span under consideration (2000-2009) because it saw a sudden deterioration of world economy with the rise of a very serious economic and financial market crisis in 2008. Based on the above, this study seeks therefore to answer the following research question: Has recourse to apologetic discourse changed in the period examined in quantitative or qualitative terms? More specifically, can the strategies classified by Ware and Linkugel (1973) be identified in company reports across the time span considered (i.e. before the crisis and after its inception)? What is their discursive realization, and does it vary over the period considered? If variations are found, have they had any impact on the generic integrity (Bhatia 2004: 112ff.) of the reports?

1.2. Corpus description

The analysis presented in this chapter is based on a corpus of company reports, retrieved from online sources, issued by UK companies that are currently members of the FT30 index, an index based on the share price of 30 British companies from a wide range of industries. It was established in 1935 and is the oldest continuous index in the UK and one of the oldest in the world. The reason for choosing the FT30 was that its stability allowed a diachronic perspective. In fact, 17 out of 30 of the companies listed in 2000 were still included in the index in 2009, therefore data from both years could be compared.

As far as corpus retrieval is concerned, both Chairman's and CEO's Statements were easily detachable from the rest of the annual report: the widespread practice of publishing company reports on corporate websites has further enhanced their hypertext-like structure, and companies seem to capitalize on sub-genre mix in ways that help offset negative information. Despite the convenience of the electronic format, manual processing and traditional qualitative investigation were necessary in order to conduct a proper analysis of the rhetorical strategies apologetic discourse relies on, as they are typically context-dependent. Qualitative analysis was useful especially at an early stage of this research project, in which potential lexical markers of apologetic discourse were yet to be identified.

Directors' Reports contain details which are mandatory in the UK, while in CEOs' and Chairmen's Reports companies voluntarily provide additional information. The latter are not only the privileged places for conveying new and forward-looking information – which readers should otherwise infer from financial statements – but they may be expected to contain 'oriented' information designed to elicit positive evaluation on the part of readers. They are a real arena for negotiating rapport with the intended readership and therefore display – where necessary – strategies to explain, promote and justify company's behavior. At this point some additional remarks on the corpus considered for analysis are needed. While it was originally planned to use the reports issued by the 30 companies included in the FT30 index in 2000, it was decided to discard 13 of them that were not included any more in 2009. Among those which were still included (17), one fund and three insurance companies have been discarded due to the different accounting standards of this kind of companies; three companies changed their names and probably underwent a radical reorganization (i.e. M&A); one, Diageo, does not provide annual reports from its archive unless you are a shareholder.

Therefore the corpus, which is about 50,000 words in size, includes reviews by both Chairmen and CEOs contained in eighteen annual company reports issued in 2000 and 2009 by 9 companies: British Airways (BA), British Petroleum (BP), British Telecom (BT), GKN (GKN), Invensys (INV), Marks & Spencer (M&S), Tate & Lyle (T&L), Tesco (TES), Vodafone (VOD).

It is to be noted that in 2009 two companies (Tate & Lyle and Tesco) included in the corpus performed quite successfully despite the economic downturn: this fact actually affected the use of the four strategies of apologetic discourse, as will be demonstrated in the chapter.

2. Narratives

The usefulness of corporate narratives to guide the reader through the meanders of financial information has long been recognized. In 1973 the International Accounting Standards Committee[3] was established to harmonize accounting standards throughout European companies. It was clear from the very beginning that bottom line figures alone could not help readers understand companies' performance. Thirty-two years later, in 2005, the International Accounting Standards Board published *Management Commentary*, a discussion paper on this subject and in February 2009 the Financial Crisis Advisory Group was still discussing the topic. Authors of company reports are aware that narration is both a central component in human interaction and one of the most ubiquitous and powerful discourse forms: it is read faster, remembered better, and comprehended more easily (Jameson 2000: 8-12). It can also be more convincing than numbers: *In Charisma We Trust* (Fanelli/Misangyi/Tosi 2009) is the title of a study carried out by an international association for practitioners in the fields of operations research and management science that seems to confirm that the way financial news is delivered can mislead even analysts' ratings. On the one hand when dealing with raw data accountants actually make their

3 IASC, renamed International Accounting Standards Board (IASB) in 2001.

own choices,[4] on the other hand, the company's management also make their choices and inevitably confer a positive spin on the story they present to the readers.

The potentiality of corporate narratives to influence readers' opinions has been widely comprehended by authors of company reports who, in fact, have extended and made narrative disclosures more sophisticated in recent years. According to a survey by Arthur Andersen (2000) of 100 listed UK companies, narrative materials occupied 57% of the annual report in 2000 as compared with 45% in 1996, just four years before. Lee (1994) found clear trends in company reports: voluntary disclosures increased faster than mandatory material. The *corpus* selected for the present study is no exception. "Previous communication tools have been put on steroids" said one of the financial communicators from a focus group with 13 communicators responsible for financial communication for banking and financial institutions in the US (DiStaso 2008).

Why are companies spending so much time and money on narratives and on company reports in general? Because company reports are the most important tool in corporate communication and if a company wants to engage in apology, that is the right moment to do so. Companies want to tell a happy ending story regardless of financial results. We never read about serious, unsolvable situations because if the situation is really serious, companies simply cease telling the story. When a company is involved in bankruptcy proceedings, for example, it usually stops issuing reports since they are not mandatory any more. Readers are therefore aware of the overall positive message companies intend to give in annual reports and of the tendency to emphasize good news over bad news, and will expect corporate information to be provided in a manner which sheds the best possible light on the company.

4 See for example the opening paragraph of the 'Notes of the Group financial statement' in Tesco 2009 Company Report: "the preparation of the consolidate financial statements requires management to make judgments, estimates and assumptions" (TES 2009: 72).

3. Deployment of strategies in ACRs

As highlighted before, the rationale underlying this study is that apologetic discourse in ACRs is expected to vary in time, also as a function of companies' economic performance. Understanding how rhetorical strategies change over time in response to changes in external circumstances is the main focus of this study. Therefore a preliminary assessment of financial data was necessary.

The financial statements of all the companies that make up the corpus have been first analyzed to verify how companies had performed in the previous year. In fact, data can be really ambiguous, that is why it was also necessary to consult expert informants from corporate restructuring studies to obtain an independent evaluation of economic performance. The main point was to understand if the companies under consideration were performing strongly or poorly relatively to the previous year. Table 1 shows an evaluation of the companies' performance in 2000 and in 2009 in comparison with their performance the previous year (i.e. 1999 and 2008 respectively). In one case, Tate&Lyle, expert informants could not assess company's economic performance univocally.

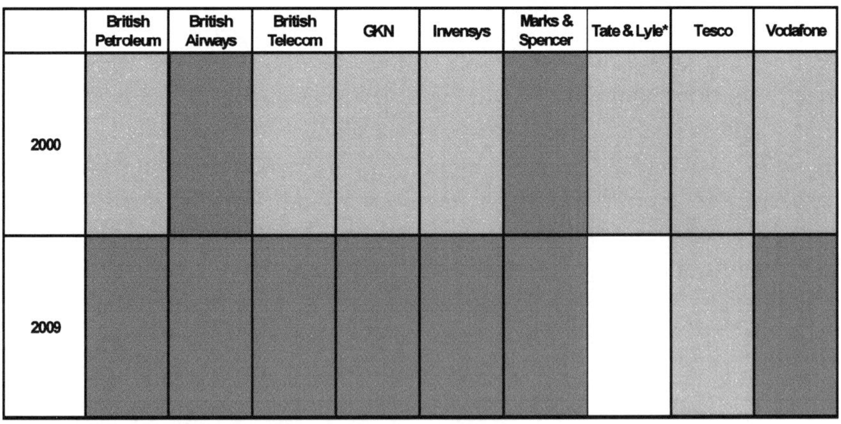

Table 1. Worse (*dark gray*) / better (*light gray*) economic performance.

The distribution of the four strategies in the 18 ACRs considered in the study are illustrated here below.

	British Petroleum	British Airways	British Telecom	GKN	Invensys	Marks & Spencer	Tate & Lyle*	Tesco	Vodafone
2000		T					T		
		De		De *(once)*			De		
		DI							
	B	B	B	B	B	B	B	B	B
2009	T	T	T	T	T	T	T		T
	De	De	De	De	De	De	De		De
		DI	DI	DI	DI	DI			DI
	B	B	B	B	B	B	B	B	B

Table 2. Strategies distribution (key to symbols: T =Transcendence; De = Denial; Di = Differentiation; B= Bolstering)

Generally speaking when companies use transcendence they intend – as mentioned above – to reframe the situation in a broader context: in company reports issued in 2009 the broader context was typically the gloomy macro-economic scenario, i.e. the financial and economic crisis, which is usually described at the very beginning of both the Chairman's and the CEO's Reviews. For example in British Airlines' 2009 company report, at the very beginning of the Chairman's Statement a whole page is filled with the following words written in giant characters:

(1) We are in the grip of a devastating global economic downturn and the next year will be extremely difficult for us. In times of crisis, it is essential that we keep our sights fixed on the future (BA 2009: 4).

The Chairman's Statement that follows fills three pages where you can find reference to "the pace of economic slowdown", the "sharp fall in premium traffic", the "record oil prices and adverse current movements", the "very tough trading environment", "the collapse of the banking system", "the credit squeeze on business and homeowners", "rising unemployment". The CEO's Review is similar in style and, again, we find preliminary remarks on a separate page, to place greater emphasis on them: "The airline industry is enduring the toughest times in its history and we expect more pain before things improve" (BA 2009: 4). This is the opening of the CEO's Review:

(2) It is hard to exaggerate the severity of the current economic conditions. During the year the credit crisis spread quickly from the US to the rest of the world provoking a global downturn that has left no single region and very few business sectors untouched. Added to that we saw that the value of sterling plunged; a collapse in consumer confidence; the continuing suffocation of credit markets; record of oil prices [...]. Airlines across the world are being battered by this storm. We have seen some 35 carriers go out of business or forced into rushed mergers [...]. Meanwhile the competition remains fierce, particularly at Heathrow and on important transatlantic routes (BA 2009: 4).

The economic scenario was radically different in 2000 and transcendence implied references to company-related crisis and not financial and economic crisis. The example which follows is taken from British Petroleum 2000 CEO's Review:

(3) In the oil sector, prices were 15% down against the level seen in the previous year; refinery margins were weak; retailing was fiercely competitive; and in the chemicals sector margins were at levels below those seen at the bottom of the previous business cycle [...]. Our chemicals business was restructured in the face of a sharp sectoral downturn (BP 2000: 12).

At British Airways both the chairman and the CEO mainly refer to competition, and at Tate & Lyle they talk about over-capacity in the US sugar market and adverse, difficult market conditions. No traces of transcendence were found in the remaining company reports issued in 2000.

Differentiation is typically used to turn readers' attention on positive data and divert them from negative figures. For example GKN, the automotive industry, in its 2009 company report lays emphasis on shareholders' support and the cutting down of interest costs, while they do not mention the reasons why they needed a rights issue.[5]

---

5 A rights issue is an option that a company can opt for to raise capital under a secondary market offering or seasoned equity offering of shares to raise money. Companies usually opt for a rights issue either when having problems raising capital through traditional means or to avoid interest charges on loans.

(4) The rights issue launched in June was strongly supported by shareholders, with over 95% acceptances, and raised £423 million before expenses. [...] we have been able to repay our revolving credit facilities and buy back £124 million of our 2012 bound, thereby reducing interest costs going forward (GNK 2009: 3).

British Airways 2009 company report also gives a very good example of differentiation:

(5) Financially, this is a very difficult time for our business. Operationally however, we are achieving levels of excellence today never seen before in this airline. [...] Shortlanded baggage at Heathrow in March 2009 was 19.2 bags per 1,000 passenger, 72.3 per cent better than March 2008 (BA 2009: 3).

Focusing on details concerning shortlanded baggage is a way to distract the reader's attention from the company's poor financial performance. These words are part of the Chairman's Statement the reader can find in the following pages and they are quoted also at the very beginning of the company report. They occupy a prominent position in the text and are a verbal counterpart to some graphics dealing with the same topic.

In 2009, Marks & Spencer performed badly and exploited, among other strategies, the indirect denial strategy, to respond to charges that are not formulated explicitly, but are nonetheless implicitly acknowledged by the company:

(6) We are now increasing the pace of change and operational execution in the business and reinvigorating our brand communications [...]. During the year we acted decisively to meet the challenges of the global economic downturn, taking steps to manage costs tightly and respond quickly to the changing needs of our customers. [...] We are now back on track after a challenging period. [...] Over the past five years we have invested heavily in the business and re-established our value credentials. As a result we have grown market share by both value and volume, and our brand is back in the hearts and minds of our customers (M&S 2009: 2).

The following is an example of indirect denial from British Telecom CEO's Statement in 2009; please note that it is also the conclusion of the overview:

(7) So in the past year we have taken some tough decisions and we have taken some bold decisions too […], but I believe the actions we have taken will enable BT to come out of this recession a stronger and better business (BT 2009: 4).

In the corpus analysed in this study, the indirect denial strategy as well as transcendence and differentiation are consistently used in the company reports of companies performing poorly, both in 2000 and 2009. Table 3 combines the presence of the four strategies with companies' economic performance.

	British Petroleum	British Airways	British Telecom	GKN	Invensys	Marks & Spencer	Tate & Lyle	Tesco	Vodafone
2000	B	T De Di B	B	De *(once)* B	B	B	T De B	B	B
2009	T De B	T De Di B	T De Di B	T De Di B	T De Di B	T De Di B	T De B	B	T De Di B

Table 3. Strategies and performance (key to symbols: T =Transcendence; De = Denial; Di = Differentiation; B= Bolstering)

In the next paragraph the variation in strategy use identified above will be illustrated through reference to the narratives included in the annual reports issued by Vodafone in the two years considered. The aim of the analysis is to highlight the way in which the shift from a relatively stable economic outlook to a crisis scenario has impacted on the discursive structuring of annual report narratives, bringing about changes which appear to have steered the rhetorical organization of annual reports towards a more extensive exploitation of apology-related strategies, thereby possibly affecting the overall integrity of this genre.

3.1. Case study: Vodafone Group Plc

Vodafone 2000 and 2009 annual company reports give the reader an excellent example of the tendency identified in the previous sections: apologetic discourse strategies are deployed, or not deployed, depending on the company's financial performance. In 2000 the company posted the best results ever, as the chairman underlines at the very beginning of his statement: "It gives me great pleasure to report to you that the financial year ended March 2000 has been the most momentous, exciting and successful yet in the history of your Company" (VOD 2000: 7). There is no trace of transcendence, indirect denial or differentiation either in the Chairman's or in the Chief Executive's Statements in the 2000 Report. Supported by fabulous financial results and strong numbers, the company ventures to present the year's results in a quite unorthodox way: the Chief Executive's message is conceived as an interview with Sir David Frost, a famous journalist and media personality. This part of the report, together with the Chairman's Statement, is separated from the rest and can be of great impact on readers, who are used to traditional chairman's letters.

As stated before in this study, bolstering is the only strategy that is always present, so it comes as no surprise that it is the only strategy that it is possible to identify in Vodafone 2000 annual report. More specifically, the strategy is employed in a part of the interview with the chief executive where the social responsibility topic is tackled:

(8) I think we take our responsibilities to society and community very seriously. We already have our own charitable activity, but it's broader than that. We have commitments to the community in many of our countries where we undertake things which we think are helpful and beneficial to the community, not with a cash or a direct reward motivation but because we believe in the community as a whole. We're also very conscious of our responsibilities to the environment and we have programmes to reduce the impact of our transmission equipment by blending it into the background as much as possible [...]. (VOD 2000: 4-5)

Nine yeas later, in a completely different macro-economic scenario, the company still engages in the strategy with a longer, more detailed paragraph, strategically positioned at the end of the report:

(9) Finally, our Board has continued to fund the work of The Vodafone Foundation, which is an important way of supporting the communities and societies where we make our profits. We invested £48 million in The Vodafone Foundation programmes during the 2009 financial year. The Vodafone Foundation and the network of national affiliates in our markets continue to achieve high recognition for the contribution they make. Highlights from the Vodafone Foundation programme over the past year include World of Difference, which helped individuals from 12 of our markets to take a year to work for the charity of their choice; a public health mobile data gathering system ('episurveyor') helping to prevent the spread of disease in 22 African countries; and the Health Alliance, announced in February 2009 with the Rockefeller Foundation, which will promote the use of mobile technology in finding solutions to healthcare challenges (VOD 2009: 3).

Vodafone's financial performance in 2009 deserves a few words. In the present study it has been assessed as "worse than the previous year" (see Table 1). Even if revenue and adjusted operating profits are higher than in 2008, other indexes need to be considered to get the whole picture. The company remained a strong company in terms of profitability, but the new high-risk context is reflected in share price decrease. The management is strongly aware of the implications of such a decrease and in the very first paragraph they engage in transcendence: "Regrettably, the share price has declined by 17% since the beginning of the year, from 154.3 pence to 127.5 pence, but has nonetheless outperformed the FTSE100 which has declined by 24% over the same period" (VOD 2009: 2). To enhance the message, drafters also resort to visuals, providing the reader with a valid instance of visual realization of the strategy: a coloured table in the second page of Chairman's Statement fills one third of the page and is aimed at comparing Vodafone's share price trend with the FTSE 100 share price trend (Vodafone's share price +7% vs. FTSE 100). The transcendence strategy is deployed here and there in the text, from the very first paragraph ("pressure on revenue in challenging economic circumstances") to the very last ("The Board is confident that the Company will be well positioned for economic recovery when it comes"). In ten out of thirteen paragraphs the strategy is present, with continuous reference to the "rapid economic decline", the "economic crisis", the "economic recession", etc.

The remaining two strategies of apologetic discourse are present in Vodafone's 2009 ACR as well. The clearest example of differentiation is provided at the beginning of the sixth paragraph, when the reader's attention is turned to Vodafone's money transfer system, after the chairman announced the reduction in operating costs that 'sadly' involves reducing the company's workforce:

(10) Your Company will continue to promote innovation in products and services across the range of our markets. For example, over 6 million people are now using the Vodafone Money Transfer system (branded M-PESA in Kenya) in Kenya, Tanzania and Afghanistan. In total, they are sending approximately US$200 million a month, mostly as small transactions of less than US$20 (VOD 2009: 2).

Focusing on minor lines of business is fairly frequent on the part of ACR drafters when they engage in differentiation.

The indirect denial strategy, a response to possible charges, is not so evident as the other three strategies in the ACR under consideration, though it is present in the tenth paragraph:

(11) Your Board refreshed the Company's strategy in November 2008 and set strategic priorities which it believes will help your Company come through the economic crisis. The review [...] put renewed emphasis on operational performance, tight control of costs, free cash flow generation and a cautious approach to further footprint expansions (VOD 2009: 3).

As stated before, Vodafone's remained a profitable company in 2009 but performed worse than the previous year. If on the one hand poor performance is revealed to the reader by numbers (share price), on the other hand it is confirmed by the use of all the four strategies of apologetic discourse. The case study under consideration is no exception in the corpus analysed in this work: when companies perform poorly they use all four strategies, otherwise bolstering is the only strategy to be present in times of plenty. For readers of ACRs, words may be more revealing than detailed financial data and sophisticated, multicoloured graphics.

4. Conclusions

The foregoing analysis has shown that apologetic strategies can be detected in corporate narratives both at times of crisis (as was expected) and during periods of economic stability, albeit with significant differences. This suggests that some forms of apologetic discourse – most notably bolstering – are part of ACRs' distinctive generic resources, as they are associated with companies' needs to account for their strategic choices and actual performance. The presence of bolstering in all the texts examined indicates that its use is part of the integrity of this genre, rather than a resource exceptionally deployed only under specific circumstances threatening such integrity.

However, apologetic discourse is not always used in the same way, rather it changes in different contexts and market conditions. When companies perform poorly, they tend to use all four strategies of apologetic discourse:[6] transcendence, indirect denial, differentiation, and bolstering. Of course, in the reports analysed this happened more frequently in 2009 due to the financial crisis. At the other end of the scale, when companies perform strongly they do not need to use all the four strategies. This suggests that apologetic discourse is indeed deployed more frequently at times of crisis, although it does make an appearance also in times of plenty; in other words it is used not only when downturns – big or small – are faced, but also under relatively unchallenging circumstances, thus confirming CEO's letter to be a "highly rhetorical product" (Hyland 1998: 224). Of these strategies, bolstering is the only one that is always present, being aimed at enhancing company image by linking it to abstract values. Bolstering is ubiquitous, and recourse to it does not seem to be dependent on the company's financial performance.

The increase in bolstering strategies linked to abstract values detected in Vodafone's annual reports is particularly interesting, as it seems to indicate a development in the rhetorical strategies of self-presentation deployed in annual report narratives. If an insistence on

6 With only one exception: Marks & Spencer 2000 Company Report, where no traces of transcendence and differentiation were found.

corporate citizenship seems to be part of a broader trend, a focus on values which are not strictly related to financial performance can also be interpreted as a way to divert attention from less-than-satisfactory market results. Thus, it may be argued that the market crisis has enhanced an existing trend in corporate communication strategies by encouraging companies to select rhetorical tools suitable to deployment in times of economic downturn, thereby acting as a catalyst in the generic evolution of corporate narratives.

SYLVAIN DIELTJENS / PRISCILLA HEYNDERICKX

The CEO's New Year's Message from the 1960s to 2000

1. Introduction

In an earlier study (Heynderickx/Dieltjens 2000) we analysed New Year's messages of CEOs of Belgian companies. Those messages are addressed to the employees and traditionally appear in the December or January issue of the corporate magazine. For this study we have collected historical material in the archives of different Belgian companies in order to analyse the diachronic evolution of the CEO's New Year's message. We have combined quantitative and qualitative methods (content, structure, language, style). The results of this discourse analytical approach reveal how the New Year's message has changed in the last few decades of the twentieth century. At the same time we have compared New Year's messages written in times of crisis with those written in more stable times.

2. Background, aim and method

Our 2000 study revealed distinctive characteristics with regard to content, language and graphic design. The messages are a mixture of business-related announcements (e.g. profit-loss figures) and more 'social' or personal remarks. At the same time, good news and bad news alternate. Most paragraphs are structured in such a way that the good news seems to buffer the bad tidings. On the discourse level, our analysis revealed a preference for the sender's perspective ('I', 'the board', 'management'). This is surprising in messages that one would expect to be receiver-oriented since the readers are offered best wishes. The texts are stereotypical and use recurring collocational phrases and clichés. The text is presented as a letter and the CEO is portrayed in his working environment. The paragraph with the best wishes is often separated from the rest of the message by means of graphic marking (italics, bold print etc.). In short, the New Year's messages of CEOs are a mixture of different genres: they contain elements of letters, greeting cards and annual balance sheets.

Literature related to the subject is that on CEOs' letters (also referred to as Letters to Shareholders/Shareowners/Stockholders or Executive Letters, cf. Garzone 2005b), since more than half of the messages have the format of a letter (see below). A search in the online databases Academic Search Premier, EBSCO and Web of Knowledge[1] resulted in eight articles focusing on different aspects and using different methods. For example, Hyland (1998) analyses meta-discourse in CEOs' letters; Amernic *et al.* (2007) concentrate on metaphors. As far as methodology is concerned, both linguistic analysis (e.g. Hyland 1998) and content/thematic analysis (e.g. Conaway/Wardrope 2010) are used. CEOs' letters have not only been examined in English but also in other languages, e.g. French (Bournois/Point 2006). As far as we can determine, none of the studies mentioned is diachronically oriented.

Specific literature on New Year's messages seems inexistent since there are no relevant references in Academic Search Premier, EBSCO and Web of Knowledge.

2.1. Corpus and methodology

We collected the corpus materials in various ways: we placed a call on the forum of the BViC (Belgian Association for Internal Communication), we contacted some companies personally and we visited the archives of several Belgian companies, mainly in the banking and the insurance sector. This resulted in a corpus of 142

1 If not otherwise indicated all websites were visited in September 2010.

New Year's messages (46,865 words) from the sixties to the late nineties (cf. Table 1).[2]

Company	*Number of messages*	*Words*
AG insurance	11	2,382
Atlas Copco	19	9,014
BAC	11	5,612
De Post	28	6,405
Fortis	4	2,222
Generale Bank	20	7,634
Kas Echo's	13	5,223
Krediet aan de Nijverheid	9	3,353
Kredietbank	17	2,293
Nationale Bank	10	2,727
TOTAL	142	46,865

Table 1. The corpus material.

We analysed several aspects of the messages, e.g. length, format, proportion of negative vs. positive information, structure of the message and linguistic features. Several aspects were determined by manual counting, since no digital versions of the old texts are available. We carried out a minute text analysis. Whether the method used was quantitative or qualitative, the aim was to pinpoint differences between the older and the more recent New Year's messages.

First, we will give a brief historical overview of the corporate magazine in Belgium. Then we will discuss some quantitative and qualitative aspects of the corpus material.

2.2. Corporate magazines in Belgium

Since we have made a diachronic analysis of New Year's messages, which are traditionally published in corporate magazines[3], we will

2 The earliest New Year's message in our material was published in the KB Bulletin of 1939-1940. It was, however, not written by the CEO but by the editorial staff of the magazine. The oldest message written by a CEO was published in the same magazine in December 1948.

start by giving a brief historical overview of those publications. The first corporate magazine in Europe, *Fabrieksbode* [Factory News], was published in the Netherlands in 1882. The first Belgian corporate magazine, published in 1920 by Bell Telephone Mfg in Antwerp, was called *Arbeid en Ontspanning* [Work and Leisure]. By the beginning of the Second World War, twelve Belgian companies offered a corporate magazine to their employees. After 1950 the number of corporate magazines increased. According to a 1971 study of corporate magazines by the Federation of European Industrial Editors Association (FEIEA), there were at that time eighty such magazines in Belgium and Luxemburg, with a print run of 300,000 copies.

3. Content and length of the New Year's messages

We started by characterizing the New Year's messages in a general way. Most of the messages (111/142, 78.17%) are a mixture of wishes and business-related information; less than a fifth of the messages (27/142, 19.01%) consist exclusively of wishes (cf. example 1). The remaining three messages contain no wishes, although they are labelled as New Year's messages. As Table 2 shows, the majority of messages that consist exclusively of wishes, dates from before 1975.

(1) Als bij een toneelstuk gaat het doek op voor het eerste bedrijf: 1988 begint. Als wij ons op onze levensscène inzetten voor ons werk, ons gezin, onze hobby, dan wordt 1988 een geslaagd jaar. Mijn beste wensen van geluk en voorspoed. (De Post, 1987-1988) [As in a play, the curtain rises for the first act: 1988 begins. If we continue to dedicate ourselves on the stage of our lives to our jobs, our families, our hobbies, then 1988 will be a successful year. My best wishes for happiness and welfare.] (All translations are ours)

---

3 Corporate magazines are also referred to as company magazines, company publications, staff magazines, employee house organs, employee magazines, employee publications, house publications, house organs, house magazines, industrial magazines, firm magazines, plant publications, factory newspapers or work journals.

Period	*Business*	*Wishes*	*Business + wishes*
1960-1964	0	6	4
1965-1969	0	7	12
1970-1974	0	4	17
1975-1979	3	2	20
1980-1984	0	2	19
1985-1989	0	5[4]	19
1990-1994	0	1	11
1995-1999	0	0	9
TOTAL	3	27	111

Table 2. General characterization of the messages.

A striking evolution in the New Year's messages is that they became longer over the years (Table 3). At the same time, the number of words dedicated to the actual wishes decreased (Table 4). Example 2 contains a fragment of text we label as a wish.

(2) Ter gelegenheid van het nieuwe jaar bied ik aan alle personeelsleden mij beste wensen van geluk, voorspoed en gezondheid aan. Moge 1985 voor U een vruchtbaar, vreugdevol en broederlijk tijdperk wezen. (De Post, 1984-1985) [On the occasion of the New Year I offer to all employees my best wishes of happiness, welfare and good health. May 1985 be a fruitful, joyful and brotherly era for you.]

In other words, the amount of business-related information increases. The business-related material contains both a look back at the year that has passed (e.g. balance sheet, events) and a look ahead at the upcoming year (e.g. financial or economic prognosis, new projects).

4 Four of these five messages come from the same company.

Decade	*Average length (number of words)*
1960-1969	213.48
1970-1979	349.61
1980-1989	324.64
1990-1999	416.5

Table 3. Average length of the messages.

Decade	*Total words*	*Total words wishes*	*%*
1960-1969	6,191	1,281	20.69
1970-1979	15,383	1,869	12.15
1980-1989	14,609	1,775	15.15
1990-1999	9,996	686	6.86

Table 4. Wishes component.

Another recurring element of the text that is on the wane over the years is the expression of gratitude. Table 5 shows how many messages in the different decades contain words of gratitude from the CEO for his personnel (example 3).

(3) Ik stel er prijs op ieder afzonderlijk te danken voor zijn aandeel in het gemeenschappelijke werk, voor de nauwgezetheid en de toewijding waarmee hij zijn taak volbracht. (Nationale Bank 1961-1962) [I want to thank everyone personally for their contribution to the joint effort, for the meticulousness and the dedication with which they have performed their duty.]

Period	*Messages with words of gratitude*	*Percentage*
1960-1969	13/29	44.83%
1970-1979	16/45	35.55%
1980-1989	12/46	26.09%
1990-1999	3/22	13.64%
TOTAL	44/142	30.99%

Table 5. Words of gratitude.

On the other hand, there is a slight increase in the number of messages that contain phrases stating that the CEO is counting on his employees (Table 6, example 4).

(4) Onze onderneming bekleedt een belangrijke plaats op de markt. Het is absoluut noodzakelijk dat dit de komende jaren ook zo blijft. Daartoe moeten wij kunnen rekenen op uw medewerking, uw wilskracht en uw doorzettingsvermogen. (AG, Forum 1996/1). [Our company holds an important position in the market. It is of the utmost importance that this remains unchanged during the coming years. To achieve this, we need to be able to count on your cooperation, your will power and your perseverance.]

Period	*Messages with 'we are counting on you'*	*Percentage*
1960-1969	4/29	13.79%
1970-1979	8/45	17.78%
1980-1989	7/46	15.22%
1990-1999	4/22	18.18%
TOTAL	23/142	16.2%

Table 6. 'We are counting on you'.

In our 2000 research project we noticed that the New Year's messages are a mixture of positive and negative information. We examined the same aspect in the material used in this study. We divided the texts in chunks of 50 words and labelled each chunk as either positive (example 5), negative (example 6) or neutral (example 7). Words of gratitude were for example labelled as positive information, fragments describing the economic or financial crises as negative.

(5) Er is een *hechtere band* ontstaan tussen deze twee groepen medewerkers van de Nationale Maatschappij voor Krediet aan de Nijverheid. Hiervoor alleen al is het jaar 1987 voor mij *een prachtig jaar* geworden. Een *steviger basis* om het jaar 1988 samen aan te vatten kan ik mij niet indenken. (NMKN, Informatie 1987/4 – our italics) [A *closer bond* has developed between those two groups of employees of the Nationale Maatschappij voor Krediet aan de Nijverheid. That alone has made 1987 *a wonderful year* for me. I cannot imagine *a more solid base* to start off the year 1988 together.]

(6) Maar met *het voortduren van de crisis*, kunnen het *verdwijnen* van ondernemingen, het *tekort* op de betalingsbalans, de zeer *onregelmatige* rentevoeten en deviezenkoersen alleen maar *een nadelige invloed* uitoefenen op de evolutie van haar resultaten, op de zo noodzakelijke groei van haar eigen middelen die haar expansie moeten waarborgen, en ten slotte – God verhoede het – op de tewerkstelling. (Generale Bank, G-magazine, 1980/12 –

our italics) [But as *the crisis lingers on*, the *disappearance* of companies, the *deficit* in the balance of payments, the *very erratic* interest rates and exchange rates can only have a *harmful* influence on its proper means that have to ensure its growth, and finally – God forbid – on employment.]

(7) Het grondstuk voor de sporthal is gelegen op de hoek van de Folklorestraat en de Kleine Doornstraat en zal toegankelijk zijn langs het sportcentrum. (Atlas Copco, Airpower Nieuws 1981/9). [The piece of land for the sports hall is situated on the corner of the Folklorestraat and the Kleine Doornstraat and will be accessible through the sports centre.]

Table 7 contains an overview of the results. Over the years the amount of negative information has increased significantly (10.3% → 29.1%), while the positive news continues to take up approximately half of the text (53.7% ≈ 50.3%). Especially remarkable is the large amount of negative information (42.45%) in the years 1990-1994, for which we have no clear explanation.

Period	*Number of chunks*	*Positive*	*Negative*	*Wishes and neutral*
1960-1964	46	24 (52.17%)	3 (6.52%)	19 (41.3%)
1965-1969	90	49 (54.44%)	11 (12.22%)	30 (33.33%)
1960-1969	136	73 (53.68%)	14 (10.29%)	49 (36.03%)
1970-1974	126	74 (58.73%)	18 (14.29%)	34 (26.98%)
1975-1979	165	95 (57.58%)	31 (18.79%)	39 (23.64%
1970-1979	291	169 (58.08%)	49 (16.84%)	73 (25.09%)
1980-1984	146	71 (48.63%)	37 (25.34%)	38 (26.03%)
1985-1989	135	86 (63.7%)	17 (12.59%)	32 (23.7%)
1980-1989	281	157 (55.87%)	54 (19.22%)	70 (24.91%)
1990-1994	106	41 (38.68%)	45 (42.45%)	20 (18.87%)
1995-1999	69	47 (68.12%)	6 (8.7%)	15 (21.74%)
1990-1999	175	88 (50.29%)	51 (29.14%)	35 (20%)
TOTAL	883	487 (55.15%)	168 (19.03%)	227 (25.71%)

Table 7. Distribution of positive and negative information.

3.1. Format and graphic aspects of the New Year's messages

We have also analysed the format in which the New Year's message is presented. A slight majority of the messages in our corpus is presented as a letter, including a form of address and a complimentary closing. Especially in the late sixties, early seventies and the eighties the letter format was preferred; it lost popularity, however, in the nineties (Table 8). In line with the Belgian conventions of letter writing one would expect that the letters at least mention the function of the sender and contain a signature. As can be seen in Table 8, there is no conformity in numbers. There are for example letters without a signature (1990-1994) or a mentioning of the function[5] (1965-1969). There are also texts with a signature (1960-1964) and a mentioning of the function (1975-1979).

Period	*Letter*	*Text*	*Function mentioned*	*Signature included*
1960-1964	4	6	2	10
1965-1969	19	6	9	8
1970-1974	13	8	14	8
1975-1979	8	16	20	8
1980-1984	13	9	17	6
1985-1989	14	10	17	5
1990-1994	5	7	8	0
1995-1999	3	7	5	4
TOTAL	73 (51.41%)	69 (48.6%)	92 (64.79%)	49 (34.51%)

Table 8. Format of the text (total: 142).

Since the messages are mainly business-oriented, an interesting aspect to consider is whether illustrations are used to personalize the text (e.g. a picture of the sender) or to add a more personal atmosphere (e.g. a snowy landscape, comparable to a Christmas card). The analysis shows that almost half of the messages have no illustration (68/142, 47.98%). If there is an illustration, it is most often a picture

5 This can be explained easily. Some messages are published under a heading like 'Christmas message of our CEO'. In those cases the function is introduced to the readers in a different way.

of the sender. That kind of illustration was especially popular in the second half of the sixties and the first half of the nineties, when there were more messages with a picture than messages without an illustration. There is, however, not a systematic diachronic evolution.

Period	*None*	*Picture sender*	*Atmospheric picture*	*Other*
1960-1964	7	3	0	0
1965-1969	4	10	3	2
1970-1974	10	7	2	3
1975-1979	17	2	2	4
1980-1984	9	8	3	4
1985-1989	11	5	6	3
1990-1994	3	9	0	0
1995-1999	7	2	0	1
TOTAL[6]	68	46	16	17

Table 9. Illustrations.

3.2. Structure of the New Year's messages

The New Year's messages have a standard structure with some clearly identifiable moves, which are summarized in Framework 1:

Framework 1: Standard structure of the New Year's message

a. (Header)
b. (Form of address)
c. Opening
d. Best wishes for the employee, his relatives and friends
e. (Christian) considerations on love, peace, solidarity
f. Perspectives for the company and the employees
g. Best wishes (+ words of gratitude) (+ 'we are counting on you')
h. Name/signature/function of the sender

6 The sum of the subtotals is 147 and not 142, since a few messages have more than one illustration.

We will now discuss the sections more in detail, also paying attention to the language used in them. Less than half of the messages (62/142, 43.66%) lack a header (a). The other messages have one or more headers: a general one like 'editorial', 'main article' or 'preface' (42/142, 29.85%), a heading referring to the new year like 'New Year's message', 'New Year's greetings' or 'End-of-year message' (24/142, 16.9%) or a heading referring to Christmas like 'Christmas message' or 'Christmas Wishes' (12/142, 8.45%). The reference to Christmas is, however, less popular than it seems at first sight, since eleven of the twelve references appear in texts from the same company. Looked at from a diachronic viewpoint, the most remarkable finding is that the heading 'Nieuwjaarsboodschap' [New Year's message] only appears after 1980.

In Framework 1, the form of address (b) is put between brackets, because not all messages include such an element. When it is introduced, it consists of a noun like 'vriend' [friend], 'collega' [colleague], 'medewerker' [employee] or 'lezer' [reader] combined with an adjective such as 'beste' [dear] of 'geachte' [esteemed]. The noun 'friend' in the form of address disappears after 1981-1982, the noun 'colleague' appears for the first time in 1974-1975. In 1978-1979 a distinction between male and female forms of address is introduced.

The texts open rather stereotypically with a reference to the situation (c): one year is ending, another year is beginning. The senders themselves sometimes point out the traditional nature of their message (examples 8 and 9).

(8) Ter gelegenheid van de jaarwisseling *is het de gewoonte* het voorbije jaar te overschouwen en daar de nodige bedenkingen bij te maken. (Atlas Copco, Airpower Nieuws, 1997/6 – our italics) [On the occasion of the turn of the year, *it is customary* to look back over the past year and make the necessary comments on it.]

(9) Het is voor mij *een traditie* geworden dat ik ter gelegenheid van het nieuwe jaar een boodschap tot alle medewerkers van De Post richt. (De Post 1992-1993 – our italics) [For me it has become *a tradition* to send a message to all employees of De Post on the occasion of the New Year.]

The language in the opening part of the message is formulaic and consists of clichés: 'the year XXXX is ending', 'we are on the threshold of a new year', 'time flies' etc. We have noticed little change over the years: the same clichés keep appearing in the text, as is illustrated in example 10 with two typical expressions.

(10) de balans opmaken [to take stock of]: 1970-1971, 1972-1973, 1973-1974, 1974-1975, 1975-1976, 1981-1982, 1994- 1995, 1995-1996

het jaar loopt ten einde [the year is coming to its end, has run its course]: 1964-1965, 1967-1968, 1976-1977, 1977-1978, 1979-1980, 1981-1982, 1986-1987, 1992-1993

The wishes are integrated twice (d and g) in Framework 1. Over the four decades we analysed, the end of the text is the favourite place to put the wishes. In 92 of the 142 messages the wishes conclude the text, in another 26 messages the wishes appear more than once in a text including a mentioning at the end (118/142, 83.01%)[7]. The wishes are very traditional and relate to good health and happiness. They are offered by either the CEO alone or by the CEO and the board of directors. An interesting linguistic feature of the New Year's messages is the extensive use of the first person singular pronouns (*I, me, mine*). They appear in almost three quarters of the analysed texts (100/142, 70.42%). In CEOs' letters, which are somewhat comparable to our material, for example, the *I*-form appears infrequently (Garzone 2005a). The wishes are directed not only to the employees but also to the people dear to them. Example 11 illustrates all these aspects.

(11) Mijn waardering – alsmede deze van mijn collega's in het Directiecomité voor uw inzet is groot en wij wensen van harte dat het komende jaar gezondheid en voorspoed zal brengen voor u en al degenen die u lief zijn. (AG Forum 1998-1999) [My appreciation – and that of my colleagues of the board of directors – is great and we wish wholeheartedly that the coming year will bring good health and prosperity to you and to all those dear to you.]

---

7 In 14 messages the wishes open the text, in six messages they are put in the middle part.

The wishes are a mixture of personal and professional elements (example 12). In that respect they are in line with the hybrid content of the New Year's messages as a whole (see above).

(12) Nogmaals good-luck en voor U en uw familie een Zalig Kerstfeest en een Gelukkig Nieuwjaar. 1971 wordt het jaar van de 20 miljard en de 700.000 spaarders. (BAC 1970-1971) [Once again good luck and for you and your relatives a Merry Christmas and a Happy New Year. 1971 will be the year of the 20 billion and the 700,000 saving accounts.]

In times of crisis (see below), the CEO displays an awareness of this crisis (example 13).

(13) Vooral wens ik dat, *spijts de moeilijke omstandigheden*, 1983 U en de uwen veel geluk en vreugde moge schenken. (Generale Bank, G Magazine 1982-1983 – our italics) [I particularly wish that – *in spite of the difficult circumstances* – 1983 will bring a lot of happiness and joy to you and those dear to you.]

Especially in the older messages there are references to values inspired by Christianity (e), like love, peace and solidarity. Example 14 clearly pleads for solidarity and forgiveness.

(14) Wij kennen allen collega's en vrienden die het minder goed stellen, die het zelfs heel lastig hebben of die gebukt gaan onder zedelijke of lichamelijke miseries. Kunnen wij hun niet een helpende hand toesteken? Er zijn er anderen tegenover wie onze gevoelens niet altijd van de vriendelijkste zijn geweest, met wie wij hebben getwist of over wie wij ons moeten beklagen. Zou het niet passen dat wij ter gelegenheid van Kerstmis een eerste stap zouden doen om een verzoening tot stand te brengen? (Kredietbank 1965-1966) [We all know colleagues and friends who are not doing well, who are even going through rough times or who are burdened with moral or physical worries. Can't we offer them a helping hand? There are others towards whom our feelings weren't always very friendly, with whom we have quarrelled or about whom we have complained. Would it not befit if we were to take a first step towards reconciliation on the occasion of Christmas?]

In the middle section of the New Year's messages attention is given to the perspectives for the company (f): what can be expected in the new year and what repercussions will there be for the employees (example 15)?

(15) Wel moeten we voor ogen hebben dat het nieuwe jaar moeilijk zal zijn voor de ASLK. De overschakeling naar nieuwe opdrachten als bank en als volwaardige verzekeringsinstelling, in een drukkend klimaat van economische en financiële crisis, zal van ieder van ons en van ons tesamen zin voor initiatief, inspanning en toewijding vergen. Als wij ons samen inzetten zullen wij deze hindernissen nemen. (ASLK, Info 1982/1) [Indeed, we should keep in mind that the New Year will be difficult for the ASLK. The switch to new tasks as a bank and as a fully fledged insurance company, in a burdensome climate of an economic and financial crisis, will require from each one of us and from all of us together a sense of initiative, effort and dedication. If we make an effort together, we will clear these hurdles.]

The final parts in Framework 1 – (g) Best wishes + words of gratitude and (h) Name/signature/function of the sender – were already discussed before.

While analyzing the New Year's messages, it became clear that the writers are well aware of the stereotypical, conventionalized and standardized nature of their texts. As we have illustrated before (examples 8 and 9), they sometimes mention the traditional nature of their message. Some of them try to counterbalance that by stressing their sincerity (example 16) or by indicating that offering the New Year's message brings them joy in spite of the obligatory nature of the task (example 17).

(16) Denk maar niet dat dit een obligate plichtpleging is. Deze wens is mij echt uit het hart gegrepen. (Kas Echo 1971-1972) [Don't think that this is an obligatory ritual. This wish really comes from the heart.]

(17) Bij dit jaareinde is het voor mij *een aangename plicht* enkele overwegingen betreffende onze onderneming te mogen formuleren. (Generale Bank, G Magazine 1977-1978 – our italics) [At the end of the year it is *a pleasant duty* for me to formulate some reflections on our company.]

With some writers, there is also an awareness of the hybrid content of the messages with their mixture of personal and professional information (example 18). The most striking fragment in this regard is example 19, in which already in the first sentence the author unambiguously indicates that he is aware that he is disrespecting the nature of New Year's wishes.

(18) Maar mijn boodschap is geen zakenbericht. (KB 1976-1977) [But my message is not a business message.]

(19) Het strookt niet met de gewoonte kerst- en nieuwjaarswensen te beginnen met beroepsaangelegenheden. De bedoeling is nochtans duidelijk: wij willen u geen werkelijkheidsvreemde begoochelingen voorschilderen met hol klinkende wensen. (Kas Echo's, 1975-1976) [It does not fit the tradition to start Christmas and New Year's wishes with professional matters. However, the aim is clear: we do not want to present you with unrealistic delusions with hollow wishes.]

3.3. The New Year's message in times of crisis

During the period 1960-2000 three major economic and financial crises occurred. In 1973-1974 there was the oil crisis and in 1979-1980, after the Iranian Revolution, the energy crisis hit. On Monday 19 October 1987, also known as Black Monday, the stock markets crashed causing for example the Dow Jones Industrial Average to drop by more than 20%. We examined whether those crises had an effect on the New Year's messages.

Elements gathered in the analysis indicate that in times of crisis the structure of the New Year's message is different from the standard structure outlined in Framework 1. As can be seen in framework 2 below, the middle part is more elaborate and pays more attention to economic considerations. The extra elements are put in italics.

Example 20 published in December 1974 contains all those additional elements: a retrospective (there were fewer investments), diagnostic considerations (inflation, restrictions on credit, high interest rates), a prognosis (no immediate change for the best) and plans for the next year (we are prepared).

Framework 2: The structure of the New Year's message in times of crisis

a. (Header)
b. (Form of address)
c. Opening
d. Best wishes for the employee, his relatives and his friends
e. Retrospective: balance positive/negative (the world, the company)
f. Diagnostic considerations (the world, the company)
g. Prognosis: positive/negative (the world, the company)
h. Plans for the next year (motivation, solidarity, more effort, economize etc.)
i. Best wishes (+ words of gratitude) (+ 'we are counting on you')
j. Name/signature/function of the sender

(20) En deze onmiddellijke toekomst dient zich niet zo rooskleurig aan. De huidige economische conjunctuur is in de ganse wereld erg onstabiel en voorspelt een zeer moeilijke en onzekere tijd. Enerzijds de snel voortschrijdende inflatie, die zich in België nog sterker doet gevoelen dan in de meeste andere Europese landen en onrustwekkende verhoudingen aanneemt, en anderzijds de geldschaarste veroorzaakt door de drastische kredietbeperkingen en de daarmee gepaard gaande hoge interestvoeten, scheppen een klimaat dat onvermijdelijk zeer nadelig inwerkt op de economische toestand in het algemeen. Dit alles heeft vanzelfsprekend ook een nadelige invloed op de investeringen waardoor de industriële activiteit in de ganse wereld uiterst ongunstig wordt beïnvloed. Een kentering in deze toestand en een heropflakkering van de conjunctuur liggen spijtig genoeg niet in 't onmiddellijk verschiet. Het hoeft geen betoog dat ook wij de weerslag van dit alles zullen ondervinden en wij in 1975 met een zekere teruggang van de activiteit zullen moeten rekenen. De evolutie tijdens de laatste 5 à 6 maanden liet ons evenwel toe dit te voorzien en gaf ons de gelegenheid ons erop voor te bereiden. (Atlas Copco, Airpower Nieuws, December 1974) [The immediate future doesn't reveal itself as bright. All over the world, the present economic climate is very unstable and predicts difficult and problematic times. On the one hand there is spiralling inflation that is more strongly felt in Belgium than in other European countries and that is reaching worrying proportions, and on the other hand there is the scarcity of money caused by drastic credit limits and the consequent high interest rates, are creating a climate that inevitably has a negative effect on the general economy. It goes without saying that all of this has a negative impact on investments,

> which unfavourably influences the industrial activity worldwide. Regrettably, a change in this situation and a revival of the economic climate are not on the immediate horizon. It is obvious that we too will feel the impact of all of this and we will have to take into account a certain setback in 1975. However, the evolution during the last 5 or 6 months has allowed us to anticipate this and has given us the opportunity to prepare ourselves.]

We have also analysed in more detail how the future is presented in the New Year's messages. In 93 of the 142 messages the CEO looks forward to the coming year. We have determined if the tone of the look ahead is positive, negative or a mixture of both. The majority of the messages has a positive outlook (Table 10).

Tone	*Number of messages*	*Percentage*
positive	58	62.37%
negative	19	20.43%
mixture	16	17.2%
TOTAL	93	

Table 10. Tone of the looks ahead.

In the crisis years (1973-1974, 1974-1975, 1979-1980, 1980-1981, 1987-1988) the outlook is clearly more pessimistic (Table 11). Example 6 is a perfect illustration of that pessimistic mood.

Tone	*Number of messages*	*Percentage*
positive	9	50%
negative	8	44.44%
mixture	1	5.56%
TOTAL	18	

Table 11. Tone of the looks ahead in times of crisis.

4. Conclusions

New Year's messages form a rather stable text type. They have a standard structure, a predictable content and a stereotypical language use. We have noticed some differences over the years. Messages consisting solely of wishes become rare from the eighties onwards (Table 2). Moreover, business related information increases. Although the texts become longer (Table 3), the amount of attention given to the wishes decreases (Table 4). Also, the CEOs feel less obliged to thank their employees for their work and their efforts (Table 5). On the contrary, they make stronger appeals for more dedication (Table 6). The analysis of the tone of the texts shows an increase of negative information (Table 7). That negative information replaces neutral chunks of text, while the amount of positive information remains more or less the same. New Year's messages with a letter format lose popularity in the nineties (Table 8). As far as the illustrations are concerned (Table 9), in the periods 1965-1969 and 1990-1994 a picture of the writer was popular. We did however not detect a distinct diachronic evolution in the presence and the choice of illustrations.

In the second part of our analysis we examined whether New Year's messages written in times of crisis differ from those written in times of economic stability. During crisis times the business-related sections of the messages become more elaborate. The middle section (cf. Framework 2) contains additional elements such as a retrospective, diagnostic considerations, a prognosis and plans for the next year. For example, more details are given about the (local and worldwide) causes of the crisis and about the measures that will be taken to counter the effects of the crisis.

A short-term diachronic analysis of CEOs' New Year's messages can reveal changes in the corporate – and even the social – climate. For example, the forms of address are revealing: the noun 'friend' in the address disappears in the early eighties and a distinction between male and female forms of address is introduced in 1978-1979.

Paul Gillaerts

From Job Announcements to Recruitment Advertising: The Evolution of Recruitment Ads in a Flemish Newspaper (1946-2010)

Job announcements are a well established genre in the external communication of organizations. It is crucial for organizations to get the right person in the right place; so, the genre occupies a prominent place in business communication and is almost continuously present in the interaction between an organization and its environment. Still, research into the genre proves to be very limited, both synchronically and diachronically. Even the normative literature on business communication focuses rather on resumés and letters of application and seems to take the ads themselves for granted. In education, too, the focus is on learning to write letters of applications and resumés, sometimes in combination with analyzing job ads, but not on learning to write the latter or use them strategically. Genre research (e.g. Bhatia 1993) also takes the perspective of the applicant and not that of the organization.

The research literature on job ads quite interestingly takes the perspective of current tendencies in business communication. Both Dieltjens/Heynderickx (2005) and Heynderickx/Dieltjens (2005) explore the buzz talk in job ads or concentrate on one specific aspect of job ads, like headers for example. Norlyk (2006) considers job narratives as a new hybrid genre reflecting the marketization of management and business communication, whereas van Meurs *et al.* (2005) investigate the impact of the use of English on a Dutch job site in an experimental case study and conclude that the supposed prestige- and image-enhancing effect does not materialize. All these studies are synchronic analyses of the job ad genre. Of course, since the eighties there have been some studies from the perspective of Human Resource Management and Social Sciences, very often focusing on the effects of the job ad, its place in the recruitment communication and the contents of the ad (Van Deuren/Lievens 1987/1989; Korswagen 1985; Timmerman 1992). Very often the aim of the research is to provide sound advice for practitioners in the field of recruitment.

1. A diachronic perspective

In this chapter we will take a diachronic perspective: we will describe the evolution of job ads in a Flemish newspaper, called *De Standaard,* over the last sixty years (1950-2010). Already before World War II the newspaper had a turbulent history; during the War it was given another title. In 1947 the old title reappeared. From then on the newspaper has appeared under the name *De Standaard* and has since become the most prominent and dominant newspaper in Flanders. Neither were the post-war period and the subsequent years without turbulence; in 1977 the editor of the newspaper got into financial difficulties and managed to be saved by a consortium of Flemish entrepreneurs, who founded a new company, 'De Vlaamse Uitgeversmaatschappij'.

The turbulent history of *De Standaard* is connected to the Flemish Movement in Belgium and ideologically the paper has always been the voice of Catholic Flanders, although in recent times this ideological adherence has been modified in a more professional and distanced approach (Durnez 1998, De Bens/Raeymaeckers 2010).

2. Methodology

Since Bazerman (1988), the diachronic approach has never been absent in genre theory and analysis. Still, most genre research has taken the historical perspective for granted rather than systematically exploring it. The only publications on the history of job ads I am aware of deal with Austrian job ads around 1900 (Ortner 1992, 1993, 1997,

1999); but even here, the diachronic, evolutionary aspect is absent. Recently, Berkenkotter (2008a) has made a case for the theoretical and methodological importance of qualitative historical genre analysis. Since genre is a cultural construct used to organize communication in society, it is crucial to study its evolutionary dynamics in order to get at a better understanding of genre as such.

The genre analysis presented here follows a text-driven procedure (Swales 2004: 72; cf. also Swales 1990) and combines analyses of structure, style and content in relation to an identification of purpose(s). The structural analysis is a move analysis (Swales 2004); the stylistic analysis focuses on metadiscourse (Hyland 2005b) and the content analysis distinguishes the main themes (De Groot 2008). Following Bhatia's suggestion (e.g. 2004: 22), our analysis wishes to go beyond description and attempts to offer an explanation by bringing together the textual findings and the contextual evolution in Flemish society and corporate communication. Thus, the job ad genre is 're-purposed' and 'realigned' in the genre network.

This chapter is, by necessity, exploratory. It combines different genre analytical approaches and limits itself to a selection of job ads published in only one newspaper. We have selected the ads from a random week in February every five years since the restart of the newspaper after World War II. We looked at the job ads on Thursday, Friday and the weekend in the first half of February; the number of ads found for every chosen year differed from twenty to hundreds. This discrepancy already testifies to the impressive growth and diversification (as will be discussed) of the job ad genre. As the years 1947 and 1948 were missing, we decided for practical reasons to take 1950 as a starting point, although the newspaper did reappear already in 1947. We decided, nevertheless, to incorporate job ads from 1944 to have a link with the war situation and be able to observe better the specificity of the 1950 job ads.

3. Analysis

Quite obviously, the analysis follows the chronology of the ads. For every chosen year we looked for typical instances of genre texts, focusing on shifts in structure, style, content or purpose. For the sake of clarity, we present our analysis in decades, with the exception of 1944.

3.1. 1944: the limits of war time

Lack of paper partly explains why the job ad remains short and is full of abbreviations. Job ads form a rather small part of advertising columns in the newspaper. The setting is rather informal; lower jobs for which board and lodging still formed the criteria. Moral criteria (e.g. "respectable maid") were also used; there were no education requirements. Metadiscourse is limited to moderate boosting, but the move structure already consists of four moves (see example 1): (1) the function (marked in italics), (2) the profile, (3) the offer and (4) contact information. Sender information is still lacking, but agencies are already present as intermediaries between the company and the applicant.

(1) Earning and living is enough (1944)[1]

> (1) *Maid wanted for flower shop*, (2) no skills required.
> (3) Very good board and lodging. (4) Keizershoek 78. Kontich (near Antw.) 10525.
> *Resp. maid wanted*, pref. from the country. Board and lodging, Klapdorp 61 Antwerpen 10524
> *All craftsm.* can find work for very good pay. Good board, 179 Antwerpsche stwg Brussel. M 2673

1 All job ads discussed here were originally published in Dutch, and have been translated into English by the author. The only exception is the ad in Example (5), which was originally published in English.

3.2. The fifties: the years of restoration

The most striking change in 1950 is the appearance of the company name, accompanied by a rather moderate boosting epithet like *trustworthy*. So, the company information remains quite discrete. Attitude markers also appear. For the rest, the layout changes slightly: it becomes bigger; frames are introduced. In spite of the drastically changed post-war context, the picture remains almost unchanged: few ads, few jobs. Also the requirements for the job remain limited to the basic skill required for it.

In 1955 there was some further variation in layout and typography. The metadiscourse used is still limited to boosting and attitude markers, but the boosting becomes more prominent. The most visible change in this respect is the inclusion of the name of the company which not only is mentioned, but is also qualified as "The best, most reliable, most known brand". This is a new move, which is inserted to precede what was originally move 1 (Specifying the function). As for the context, a reference to social laws appears. All in all, the job ad remains a rather modest and stable genre. This will change drastically by the end of the fifties.

3.3. The golden sixties

As regards the layout, in 1960 we notice a growth in the size of the ads, more variation and the use of bullets and subtitles. In other words, we can witness the beginning of a format. Metadiscourse still remains quite restricted, but a move structure starts to become more visible, following the format of the ad. Regarding the first move, we can see a logo showing up, and regarding the profile move more specifications are given, albeit with rather clumsy wording, viz. the obsolete use of a present participle. Regarding the context, we see more jobs and more ads, higher level jobs with education and training requirements. Extralegal advantages are being offered, e.g. "The car is being provided for by the company."

The main change during the golden sixties in Flanders is the arrival of multinationals (like Procter & Gamble), which clearly pushes the job ad to adopt a stronger format, with a more prominent use of boosters and attitude markers. In an ad of Procter & Gamble the company promotes itself as 'Worldwide known American company', "offering excellent and stable working conditions" and "extralegal advantages".

Most striking in 1965 is the appearance of pictures in job ads, sometimes, with an advertisement of the product, i.e. a piece of furniture as shown in Figure 1. The embedding of the ad in the first move introduces the use of engagement markers (*you*) in the job ad, and a beginning of argumentation:

> News for the equipment of your offices: Apeco polyform
> Advantages: rational space saving; adaptable.
> Tell us your problem already today.
> Prices, with illustrated documentation
> 'R' will be sent to you. (Figure 1)

Of course, we also witness a greater variety of sizes and layout. Stability of the job in combination with extra legal or social advantages as specifications of the offer move come into prominence.

Figure 1. The picture conquers the job ad (1965).

3.4. *The consolidating seventies*

By 1970 the job ad genre reaches its completion: illustrations, a dialogic style with the manifest presence of engagement markers and the introduction of transitions as part of the argumentation, all characterize the fully-grown genre of the job ad. Again, the job ads of multinationals take the lead and contribute to marketization, which is reflected in the dialogue approach used in many job ads at the time.

(2) The dialogue approach (1970) :

More and more young people are building their career with us.
Why don't you?
You are important to us.
So you can enjoy the best conditions

As you can see in the job ad for a secretary (Figure 2), the illustration of the typewriter dominates and is fully integrated into the ad.

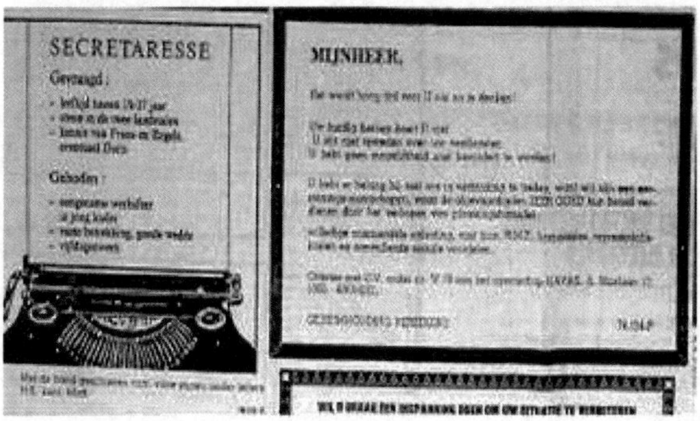

Figure 2. The prominence of the illustration and the bending of genres (1970).

The ad published next to it is interesting because it bends the genre of a sales letter. Partly as a consequence of this bending[2], we can witness the beginning of textualization: from description (with bullets, and in

2 The notion of genre bending was introduced by Bhatia (2004); he does not distinguish it very clearly from the notion of genre mixing and the notion of colonization. We use it to indicate the appropriation "of generic resources from a specific genre for the construction of another. These resources may be lexico-grammatical, rhetorical, discursive or other generic conventions, which are appropriated to give shape to more dynamic or innovative generic forms. […] These appropriations often lead to the colonization of one genre by the other by invading its integrity." (Bhatia 2004: 87) .

an elliptical style) to explanation and narration in full-fledged syntax. In the example the integration of the sales letter format in the job ad leads to more promotional and argumentative discourse. Sometimes the first move appears in a separate box and becomes an embedded genre, rather than part of the job ad.

The role of the applicant changes: s/he becomes crucial for the survival and the success of the company. Job ads gradually become a matter of recruitment communication.

3.5. The explosive eighties

By the beginning of the eighties the tendencies visible in the seventies are further confirmed. The layout has become more professional, with textual organization in paragraphs and the integration of the illustration into the job ad. Engagement markers and transitions become more important. Even a column format was introduced to associate the job ad with the news articles in the newspaper.

The most important development in the eighties is the introduction of a personnel guide in which the job ad appears. This personnel guide presents news articles on job issues and confirms the changed role of the applicant: s/he has become a client in a booming recruitment business.

Informal ads, using a spoken register and narration, are no longer unusual: the job ad has become a promotional genre. Sometimes informal language is combined with formal *you* (*u* in Dutch as opposed to informal *jij*).

(3) Even better than a secretary? Well yeah, at Allumalux, the exclusive importer of products of Bosch, Blaupunkt and Bauer… Is this possible? You [=U, PG] will be the
assistant of the head of the department of electro domestic material,
who organizes appointments and meetings, who is responsible for the correspondence of the department and the discrete confidante[3] of the big decisions. […]

---

3 Notice the overt sexism, PG.

To summarize, there are two tendencies: one in the direction of news value and journalistic genres; another in the direction of promotional genres.

3.6. The engaging nineties

In the nineties colours (blue and red) started to be used in the ads. The use of real people heightens the human interest factor of the job ad. A slogan like 'In a crowded market one name stands out: Price Waterhouse' makes clear that the focus has once again shifted to the company for which employer branding becomes an issue. The first move, sender information, dominates.

3.7. A new millennium

By the end of the century, the job ad genre shows an immense variety. The two tendencies (promotional and informational) we noticed earlier were confirmed and textualized in two subgenres: a promotional one, with dialogue, engagement markers and even rhyming (see Example 4); a more informational ('objective') one, with all five moves – including the first move, company name specification (see §3.2 above) – neatly distinguished from each other and carefully elaborated (Example 5). As can be seen in Example 5, the only one amongst those discussed here that was originally published in English, this informational format is global, as is the use of English.

(4) A Promotional Job ad of an Insurance Company (2000)

> Yes, we have jobs for dynamic graduates!
> Ready to become our new client manager?
> Does it sound familiar in your ears?
> Let's then hear from you!

(5) An Informational Job ad of an Express Company (2000)

> (Illustrations: a globe at the upper left and a logo at the bottom right)

(slogan) DISCOVER how we make it fit
(Function) *Secretary to the Finance Department*
Profile: • higher secretarial degree or equivalent experience
• computer literate: MS Word, Excel (Powerpoint and Access is a plus)
• at least 1 year of related experience • [...]
Position: • provide administrative and secretarial support to the Finance Manager and the Financial Group • assist with the preparation and distribution of monthly, quarterly and annual financial information for senior management and shareholders [...]
Offer: • a unique chance to work for an innovative high-performing and fast-growing global company • the opportunity to become part of a dynamic team dedicated to (internal) customer satisfaction through continuous development of the department's performance • [...]
(In the left margin: information about the company)
(At the bottom) If this position interests you, please send your application letter to [...]

Of course, both subgenres may be combined, as is clear from the ad for textile engineers in figure 3. All colors abound; ads may cover a whole page and deal with many jobs in one ad, implicitly confirming the importance of the company.

Figure 3. A combination of a promotional and an informational job ad (2000).

Maybe the most striking feature of these years is the break-through of recruitment companies which completely dominate the job market. In example 6 Job@ is said to offer 1,800 jobs every week. Job@ becomes a magazine, regularly included in weekend editions of newspapers, and containing pages of news, background information, thematic exploration of job related issues, etc. To help readers find their way through the enormous amount of job ads an index has been added.

(6) Advertising the recruiter (2000)

Every week choice from 1,800 jobs. It is possible!
Click on <www.jobat.be> and find the job of your life.

Interaction with other genres of job recruitment, like company open days as announced in Figure 4, company websites, recruitment company websites, etc. also prevails. All genres and media play a role in the recruitment effort to find the right man or woman for the right place. So, finally the job ad becomes scattered into a genre colony, 'a grouping of closely related genres" (Bhatia 2004: 57), which are no longer colonizing each other, but are cooperating and enforcing the effort to reach the candidate/consumer. In this respect the concept of "colonization" (see footnote 1) may be obsolete and not get the picture right.

Figure 4. Inviting applicants to an Open Day at the company (2000).

4. Discussion

The job ad genre has proved to be both stable and dynamic. It appears as a stable genre because of its early fixed move structure, gradually underlined by its bullet format. In 1944 we noticed already four moves; in 1950 all five moves were present. Still, the job ad shows great dynamism, first becoming more promotional, and then gradually expanding into other genres, at the beginning within the job ad, later also outside the job ad in different media, hence forming a genre colony. We may even consider job ads as part of a genre repertoire in which applicants are submerged by job ads, campus recruitment, job events, direct mailing, posters, radio, television, job spots, annual job guides, e-recruitment channels (commercial and official), magazines (Zobell 2002).

The basic mechanism behind the evolution of the job ad genre can be described as follows:

1. The development of the genre text (by adding moves, content, completing style).
2. A style shift from description to narration, advertising, and recruiting/selling gives way to mixing, embedding and bending genres within the job ad.
3. The explosion of the genre text and spread of the ad across different genres and media, leading to a genre colony and repertoire for job recruitment.

The job ad can be shown to have moved from the periphery in a newspaper to a central position in that newspaper, and then back into a more peripheral position again with the emergence of e-communication. Thanks to the co-existence of different job ad genres, its multifunctionality, lying basically in how it moves back and forward between promotion and information, can be guaranteed.

An analysis of the evolution of the job ad genre deepens our understanding of genre. As regards the move structure, despite its evolution, the genre shows an impressive stability in its moves, albeit with a growing variety of strategies/steps, revealing genre bending and mixing with other genres like direct mail and commercial advertising.

Metadiscourse analysis reveals the growing importance of ethos (stance), testified by the increasing use of boosting and attitude markers and pathos (interaction), shown by the increased use of engagement markers. Both are indicative of the marketization of the genre; in our content analysis new themes have been found, like extralegal forms of remuneration and ecological issues, which are immediately linked to the job seekers' interests. Finally, regarding the context of the genre, while the explosion of related (very often digital) genres within the genre system/network of recruitment marketing has pushed the traditional printed job ad to the periphery, the job ad has survived, giving birth to ever new genres that enter the centre of the genre system.

5. Conclusion

In searching for the rationale behind the evolution of the job ad genre in a Flemish newspaper, several contexts may prove instructive. The first context is political; after World War II, we witness a power shift from the Southern French-speaking part of Belgium to the Northern Dutch-speaking part. In the forties Flanders was still an underdeveloped region, devoid of industry, and remained mainly agricultural. In the fifties serious conflicts – usually centered around issues like collaboration and repression, the abdication of King Leopold III, the school war, territorial claims of the South – hinder a further development of the North. Only in the sixties, with the fixation of the linguistic border, the split of the broadcasting, the development of separate higher education and political parties, did Flanders achieve cultural emancipation. Several constitutional reforms since 1970 have led to the founding of an actual federal state, thereby empowering the regions.

The second context, an economic one, may offer an explanation for the evolution in job ads in Flanders. The decisive year is 1961 because in that year Parliament voted a so-called Unity Law, ensuring savings for the state and subsidies for international investments, which found their way to Flanders. The golden sixties witness an economic shift from the South to the North. Thanks to the democratization of education and the standardization of the Flemish variety of Dutch, Flanders became capable of creating its own economic and intellectual elite, and became one of the most prosperous regions in the world.

Thirdly, the social context may also add to a better understanding of the job ad evolution. In the fifties we witnessed a fast growing population ready to enter the job market and secured by an encompassing social security system. In the sixties the middle class became emancipated and as a result of growing participation in higher education, the development of both industry and tertiary activities become possible. All this led to the availability of a large group of applicants, who were ready and capable of taking jobs.

Fourthly, there is also a technological context, characterized by important innovations in graphics and the introduction of the computer in business communication.

Finally, the birth and the development of corporate communication is also an important factor in explaining the evolution of the job ad genre. The rise of Human Resource Management and the importance of corporate identity and image issues in a growing global market, may also partly explain the growing promotional diversification of the job ad genre.

Because of limitations in length and the lack of previous research this study is and remains exploratory. The methodology used, i.e. a genre approach integrating content/context, style and structure, has allowed us to sketch a general overview of and provide insight into the basic mechanisms involved in the evolution of the job ad genre. Of course much more empirical research will be needed. More data using other intervals are needed to refine the results. Domain-specific research (e.g. industry vs. tertiary sector) may also yield differences in evolution. Finally, intercultural (comparative) research may contribute to a better understanding of the link between the evolution of the job ad and developments in corporate communication, like the increased globalization and internationalization, for example, stemming from the dominant role played by multinationals.

FRANCESCA SANTULLI

Genre Variation and Genre Change: Theory and Applications

1. Introduction: the notion of genre

The migration of the notion of genre from the realm of literary studies to the wider area of language use has opened up new and attracting perspectives: in contemporary linguistic research genre has become one of the most widely discussed concepts, and researchers have exploited it in contexts ranging from scientific to corporate communication, from professional settings to everyday verbal interactions.

In text analysis genre is an important element to single out different registers (Halliday 1994), but it is also essential to understand the notion of intertextuality, as the latter refers not only to the insertion into a text of pieces taken from other texts and discourses (in the horizontal dimension of 'quotation'), but also to the paradigmatic dimension, entailing implicit reference to other texts, which form a recognisable cluster to which the new text ideally belongs. A genre, however, is not a finite class of texts, a discrete category whose members are recognisable thanks to a set of properties necessary and sufficient; it is rather a sort of background against which a new text is generated, which is imperceptibly modified by each new element, similar to the horizon that moves following the movement of the observer.[1] Under this perspective genre can be compared to the concept of linguistic norm (Coseriu 1974), as it bridges the gap between universal phenom-

1 The metaphor of genre as the addressee's horizon of expectation (*Erwartungshorizont*) was introduced by Jauss in his research on the emergence of new text genres in the Middle Ages, parallel to the development of new literary traditions in Romance languages (Jauss 1977). The German researcher also suggested a parallel between genre and linguistic norm.

ena and individual instances, and thus makes it easier to understand how general rules are transformed into real texts.

Considering the crucial position of genre and, at the same time, its elusive nature, it is not surprising that different definitions of genre have highlighted different aspects of the notion, which in my opinion can be usefully grouped into three main areas, focusing respectively on the goals of communication, on the expectations of the addressees or on the characteristics of texts. Obviously, these three basic aspects are not contradictory, nor do they exclude one another: rather, they represent three fundamental and complementary keys for access to communication. If we consider the metafunctions of language singled out by Halliday (1994), which are often mentioned in pragmatic analyses of texts, genre can be linked to the ideational component, as it derives from the goal of communication; it is also closely connected to the interpersonal function, as it concerns the relationship with the addressee; finally, and most obviously, genre is a textual phenomenon, as it accounts for a canon, a set of fundamental rules determining the character of a text.

It is therefore necessary to consider genre in a multi-layer perspective accounting for the presence of various aspects, which – despite their different implications – all share an important characteristic, i.e. their pragmatic nature. In other words, the notion of genre belongs to the real world of text production, and is therefore part of the history of language in its different manifestations. Genre stems from the acceptance of norms, which are often implicit but strongly embedded in the historical situation, functional to reaching the objectives of communication on the one hand, and, on the other, to meeting the expectations of the receivers. The definition of genre as a form of 'social action' (Miller 1984/1994) synthesises these different implications: it is worth considering that it also connects the modern version of the concept to its ancient predecessor, the classification of genres put forth by rhetoricians. As a matter of fact, in the classical partition originating from Aristotle's rhetoric, the three different genres corresponded to specific social situations (each having its aims and characteristics), thus emphasising what in modern terms would be called the 'pragmatic' nature of discourse.

Given the social (and thus historical) nature of genre, it is necessary to consider both its synchronic manifestations and the constant evolution of its individual forms.[2] It is impossible to explain how genres function without considering how (and possibly why) they inevitably change. If on the one hand single textual instances can be better understood against the background of shared generic norms, on the other it is necessary to bear in mind that they occur in a diachronic dimension and thus contribute to the evolution of the very norms which generated them. From this point of view it is easy to find similarities between language and genre evolution, which make it possible to apply to the diachronic analysis of genres general principles that have emerged in the long – and fruitful – tradition of studies on language change.

Against the background of these theoretical premises, the aim of this chapter is to focus on the dynamic relationship between variation and change, which was clearly described with reference to language (Weinreich/Labov/Herzog 1968) and can also be applied to genre evolution. Reference will be also made to a model that combines the notion of genre with the role of persuasion in language interactions (Halmari/Virtanen 2005a). The basic theoretical principles will be applied to a text genre of relatively recent origin: travel guides. Through the analysis of their origin and their variation, the discussion will consider the hypothesis of the existence of hybrid genres as well as the mechanisms of hybridisation, linking the function and the very existence of genres to the ultimate needs of verbal communication.

2 It is important to emphasise the 'historical' character of social phenomena to avoid the simplistic interpretation of the dichotomy between synchronic and diachronic linguistics as an opposition which excludes the former from the analysis of historical facts. Actually, the social character of language makes it a historical phenomenon in all its manifestations, both if we consider its evolution (diachrony) and if we analyse its mechanisms and its functioning in a given time-place situation (synchrony). This obviously applies also to genre analysis.

2. A model of (genre) evolution

2.1. Change and variation in language[3]

Language change was the most important topic of research for linguists in the XIX century. The study of change (above all of phonetic change) highly improved the understanding of the evolution of single languages or language families, but most researchers focused exclusively on the analysis of minute phenomena, and were not able to develop a comprehensive theory of change. The most important exception was the fundamental work by Hermann Paul (1920), in whose theoretical view language was ontologically reduced to the idiolect (*Individualsprache*); as a consequence, change in language use was simply the result of the substitution of new idiolects for older ones – a mechanism of generation turnover, which is obviously insufficient to explain why new idiolects should adopt different traits.

The problem of language change was not neglected in the development of new theoretical models: researchers in the early 20th-century, though apparently more attracted by synchronic description, applied the principles of structural analysis of language also to the diachronic dimension and tried to use them to describe and explain change. The aim of researchers was to emphasise a global approach, considering the mutual influx of changes affecting different elements at the same level of analysis – and thus extending the idea of syntagmatic 'ease' to the paradigm, in the functionalist perspective.[4] A more global approach made it possible to consider also the connections be-

---

3 Obviously, a discussion of a theory of language change lies far outside the scope of this chapter: I shall limit myself to a very synthetic presentation of my point of view on the matter, largely based on Weinreich's seminal work, emphasising the points that can be more easily extended to describe and explain genre change.

4 Martinet (1955) widely exploited the relationship among different members of the paradigm, introducing the concept of 'economy' in change. The adoption of the new theoretical approach of structuralism did not however improve actual understanding of the mechanisms which ultimately trigger change and govern its manifestations.

tween different levels of linguistic analysis, and suggest how, say, phonetic change can affect morphology and trigger a series of evolutionary processes which can lead to profound transformations in the very structure or typological nature of a language.

Despite the innovative methodology adopted by structuralism and functionalism, from the point of view of a theory of language change there was no radical improvement. Researchers still concentrated on 'inner' causes, which can describe *how* language changes, but are not capable of showing *why* change ultimately occurs. As a matter of fact it is not possible to go beyond pure description and reach the explicatory level without considering 'external' factors or reasons that lie outside language. In other words, the structuralist point of view is inherently insufficient to explain change, insofar as it considers *la langue pour soi même* – as affirmed in the famous (though probably spurious) conclusion of Saussure's *Cours de linguistique générale*.

Criticism both to 19th-century neogrammarian and 20th-century structuralist approaches to language change was the starting point of a seminal paper by Uriel Weinreich (Weinrich/Labov/Herzog 1968). Weinreich's thesis stems from a fundamental premise: language cannot be considered a rigid and homogeneous structure; rather, variation is essential to guarantee effective communication. The task of linguistic description is to consider variation and account for the distribution of the different forms which make up a language as a structured heterogeneous entity.

Once the link between structure and homogeneity is broken, it becomes evident that the existence of a system does not exclude variability. This is the core of the famous notion of diasystem put forth by Weinreich himself (1954). In a diasystem different variants co-exist and their distribution is not chaotic: geographical, social, cultural and even behavioural factors influence the actual production of speakers (who are by no means constant in their choices), and research can reveal patterns of distribution that correlate language items to genuinely non-linguistic phenomena. Variation is the pre-requisite for change, as change originates from the expansion of a variant to the detriment of other parallel forms. The reasons for change lie outside language structures and reflect other historical phenomena. These considerations, in turn, emphasise the historical (contingent and social) nature of language itself, which meets the diversified communicative needs of a given group and is therefore continuously adapted to the situations in which the speakers express themselves and interact.

2.2. From language to genre

As a purely historical phenomenon genre can be compared to language – or more precisely, as suggested above, to language norm.[5] To analyse genre in the diachronic perspective it is therefore necessary to consider its synchronic manifestations. Lack of homogeneity is common to language and genre: a genre is not defined by a closed set of rules, and texts belonging to the same genre show significant differences, which (as in language) are the premise to change.

As in the case of language change, breaking the link between genre and the homogeneity of the texts belonging to a single genre is the first step to try to understand how genres change.

The range of variation typical of a genre in a given historical moment is by no means constant: as pointed out by Halmari/Virtanen (2005b: 237), "when a genre is maximally conventionalized, *the intertextual gaps* between the representatives of that genre *are minimized*". The canon, however, can be re-interpreted less rigidly, and as a consequence the differences between texts increase: this is the process of maximizing intertextual gaps. When this flexibility leads to the production of texts that display some features typical of another genre, the interdiscursive gap between different genres is minimized: "genre boundaries are blurred and generic hybrids are born" (Halmari/Virtanen 2005b: 240).

The model described by Halmari/Virtanen is focused on the process of contamination between different genres, which is linked by the authors to the presence of a persuasive intention in texts. Apart

---

5 Here it is not possible to discuss the difference between change in language structure and change in language use (Hjelmsev 1963). However, if the comparison between language and genre takes into consideration the level of language norm, it is not necessary to go into finer distinctions, as norm is a historical phenomenon, linked to a social community, though it can by no means be reduced to the chaotic accumulation of single idiosyncratic utterances.

from this specific aspect (which will be discussed later), it is important to emphasise that the idea of genre as a prototypical category is fundamental to illustrate genre change. To be recognisable by their addressees as members of a genre, texts must share important features with their generic prototype; in certain historical conditions and for different reasons, however, they can start moving away from it, refusing some accepted elements and introducing new ones, in a dynamic pattern of variation. The result, however, is not necessarily a migration towards another genre, where texts are more easily recognisable against a different prototypical background; in my opinion, change is not always combined with hybridization (as suggested by Halmari/Virtanen). It is essential to bear in mind that the image of the original prototype is constantly and imperceptibly changed by each new text and a dramatic shift can be the result of a gradual process of accumulation.

This linear, most elementary form of evolution does not affect the system of genres as a whole: it is so obvious and inherent in text production that it is usually taken for granted and neglected in genre analysis. As in the case of language, change is not distinct from use.[6] This, in turn, is the prerequisite for more profound change, that not only affects the formal canon of the genre, but also transforms the goals of communication and can, in this process, interfere with the structures that are typical of a different – though partly similar – genre. In this case, there can be a hybrid phase in the evolution, which however ideally aims to a re-definition of boundaries and the emergence of new, stable (though constantly changing) genres.

Obviously, these consideration can be useful to describe *how* genres change. Yet they can say very little about the reasons *why* change occurs. Variation increases and emphasises the differences; one of the forms available for actual use is more and more frequently used so that the alternative variants lose ground and start to become obsolete; finally, the winning form is accepted as 'norm' causing a reassessment of the whole paradigm and possibly interference with adjacent paradigms. This is a mere *description*, definitely not an *expla-*

6 Under this perspective, the parallel with language is particularly interesting: each occurrence of the norm is the concrete manifestation of use, but – with its individual characteristics – it also contributes to the modification of the norm.

nation. As in the case of language, the reasons for change lie outside genre proper.

In the above mentioned book (Halmari/Virtani 2005a), the authors investigate change from this point of view, focusing their attention on the relationship between genre and persuasion, two key aspects of communication, which are often considered indispensable constituents in language use. Starting form the assumption that the more implicitly persuasion is codified, the more effectively messages are accepted by receivers, it is possible to infer that increased familiarity with the forms of persuasion in a given discourse corresponds to a decrease in the persuasive power of a message. "The process works in a way similar to the evolving nature of euphemisms" (Halmari/Virtani 2005b: 230). Under this perspective, persuasion is like taboo and, like the words used to euphemise unpleasant referents, its forms have to be constantly renewed. As a consequence, new linguistic – and generic – forms are continually produced.

The interesting interpretation put forth by Halmari and Virtani acquires, in my opinion, a stronger explicative power if it is considered against the background of the evolution of the specific goals of persuasion – the changes in the intentions of the addresser, who must find the right balance between overt explanation of the aims of communication and covert codification of its persuasive component.

Moreover, the aim of communication is only one of the pragmatic factors affecting change. Thinking of the tripartite classification illustrated above, it is necessary to mention at least new expectations of the receivers and the need for new textual norms, that can also be linked to new conditions for the production, transmission or storage of texts. The combination and interaction of these – and possibly other – external elements can lead to different, and in some cases extraordinary rapid forms of evolution.

Hybridization is a possibility, but it is necessary to use this word with extreme caution. As a matter of fact, all genres are hybrid, as all genres are normally generated with multiple aims to meet multiple needs, and for this reason include and skilfully combine different text types.[7] There is no text type that is exclusive of one single genre, and it is difficult to imagine a textual feature that is not shared by different genres. Hybridization occurs only when two genres merge or a genre 'colonises' its adjacent genre, possibly pushing it further and thus triggering a chain of interrelated shifts.[8] Similarly, language change does not generate hybrid languages: out of Latin different Romance languages progressively emerged and consolidated through a long process of divergence and convergence, the intertwining of Old English and Norman French gave birth to a 'new' language which displayed characteristics inherited from both its predecessors. Confusion is simply a phase in evolution.

3. Travel guides: a case study

The pattern of variation and evolution of genre described in the previous paragraphs will now be applied to the analysis of a relatively recent text genre: travel guides. Two factors will be taken into closer consideration: 1) the origin of the genre – which shows how new social needs induce a re-interpretation of previous genres and possibly a merge of different features; 2) some aspects of its recent evolution, characterized by a high level of variation and contamination with other texts (promotional brochure and reportage in the first place).

3.1. The birth of a genre

In 1832 the German publisher Baedeker gives birth to the first travel guide, dedicated to the Rhein region. Scholars have repeatedly empha-

---

7 I refer here to the classification put forth by Werlich (1975), who singles out five fundamental types: narration, description, exposition, instruction, argumentation.

8 This obviously reminds of the drag-and-push chains widely used in functionalist description of language change.

sised the importance of this publication to the point that 1832 has often been considered the 'official' date of death of the *Grand Tour*.[9] The connection of the two events stems from obvious observations: the practice of travelling across Europe (namely, through the Alps to Northern and Central Italy, possibly to Naples and even as far as Sicily) had a crucial educational role in the life of 18th-century aristocrats. It was a long demanding experience, which required long preparation as well as ample availability of time and money. The young noblemen who left the countries North of the Alps to come into direct contact with the treasuries of ancient Rome and of Renaissance had to study the fundamentals of archaeology and the history of fine arts; they needed comprehensive publications, which discussed the most important topics in depth. They also read the narrations written by previous travellers (diaries), which gave them important information about the places and the experiences they could expect.[10]

In that period, the wide production of travel books was by no means homogeneous. It is possible to distinguish at least two main areas: books mainly focused on scientific knowledge (based on geological, botanic and zoological observations on the one hand, on ethnographic analysis on the other); travel diaries, where the autobiographic experience prevailed and emphasised the literary attitude of the writer. In the latter the narrative element was predominant and reflected in the deictic organisation of the text originating from *ego, illic, tunc* [I, there, then]. Travel diaries however could also have a prescriptive component, giving suggestions to readers transformed into would-be travellers.

The decline of the *Grand Tour*, closely linked to the emergence of new social patterns, did not extinguish the desire for travelling. On the contrary, the number of people involved increased dramatically,

9 Among others, Boorstin (1986), Brilli (1995). A synthetic but precise sketch of the history of the tourist guide can be found in Ragonese (2010: 8-14).

10 A pioneer in this area was Fynes Moryson's *Itinerary* (published in 1618), which was very popular in England: it described in a series of volumes the travels of the author, whose personal experiences were considered dramatically useful to learn how to travel and to single out the most interesting sites.

but new models for the practice were adopted, more adequate to the needs and the goals of the new middle-class travellers.

This is the social context in which the new text genre developed: external factors urged a re-assessment of the previous genre system, namely in the area of travel books. The practical component had to be expanded: in this respect, the ancient tradition of 'itineraries' for pilgrims or merchants was an important source of inspiration.[11] But, apart from practical information, the reader expected description of the places and their historical background, as well as suggestions to organise the itineraries. In this new pragmatic situation, the reader is the traveller and the protagonist, here and now (*tu-hic-nunc*). The writer, with the contribution of his personal experience, disappears.

As a matter of fact, the crucial innovation in Baedeker's publication is actually the adoption of a new 'contract' with the readership: the name of the author is no longer indicated and is replaced by the publisher's, who now guarantees for the reliability of information and the advantages of the proposed itineraries.[12]

The emergence of travel guides corresponds to the drift of diaries into the realm of literature proper, with more emphasis on personal experience: the travel is often the occasion for meditation, or becomes an important step in the process of self-education, in line with some of the essential features of the *Erziehungsroman*. The guide, in turn, accepts the practical approach of itineraries, but combines it with serious artistic and historical information; it construes an image of the places to be visited which is the counterpart of the image of its addressee – the traveller and implied reader. The new pragmatic and semiotic situation corresponds to the necessity of exploiting different

---

11 The so-called *mirabilia*, Medieval manuscripts used by pilgrims bound to Rome or other holy places, can be considered "prototypes of tourist guides" (Ragonese 2010: 9).

12 Ragonese (2010) mentions a book by Giuseppe Vasi, published in Rome in 1763, that can be considered 'one of the first travel guides', as it organises the presentation of the most famous Roman monuments in eight 'sessions', corresponding to the time necessary for the reader to visit the places described in the book itself. The structure of the 'guide' is evident, and the book does not describe a personal, past experience; the author, however, is still there and is responsible for the truth of information and the effectiveness of his suggestions.

text types: description of the places, narration of their history, exposition of practical data (means of transport, timetables, conditions for visit, etc.) and – the most important and distinctive feature – instructions for the traveller. Yet this does not mean that a travel guide is a hybrid genre.[13] In order to meet complex needs, different textual strategies (inherited from different traditions) were selected and combined to give birth to a new genre characterised by a highly recognisable set of rules. In the first stages of their production – and for many decades thereafter – travel guides displayed a high level of uniformity, which extended across cultural and language boundaries, and reflected the strong similarities in the habits and objectives of (upper)middle-class travellers.

Travellers across Europe recognised the crucial role of their Baedeker's (and similar publications): the guides were handy books to be brought everywhere and guaranteed easy access to essential information and instructions. As explicitly declared in the introduction, the aim of a guide is to indicate 'privileged' itineraries, which make it possible for the traveller to see what is worthy of attention without wasting time and money.[14] Under this perspective, efficiency is a key value, and it is therefore necessary to follow instructions strictly. In the peritext the ethos of the guide is affirmed strongly and overtly, making a persuasive component evident; in the actual text, however, its voice is almost cancelled and a very transparent style is adopted, which disguises in veridictive and impersonal textual sequences the

---

13 Kerbrat (2004) considers travel guides a hybrid genre, as they combine different discourses; in my opinion it is more useful to distinguish between genres and text types, as suggested above.

14 Baedeker' *Oberitalien* (published in 1902) opened with the following words: "Das Reisehandbuch für Oberitalien will den Reisenden in den stand setzen, mit möglichst geringem Zeit- und Geldaufwand dasjenige rasch zu überblicken, was besondere Aufmerksamkeit verdient, ohne mit bedeutungslosen Einzelheiten zu überhäufen" (p. VII) [The guide book for Northen Italy will allow the traveller to see quickly - and for the smallest possible amount of time and money - what deserves special attention, avoiding the accumulation of meaningless details, *my translation*]. Similar statements can be found in all travel guides which followed the approach of the German publisher, as, for example, the red guides of the Italian Touring Club.

imposition of choices and evaluative judgements, thus imposing – literally – its 'point of view'.[15]

This aspect, which implies a covert argumentative function of the text, is another important feature of travel guides, which has also had a significant influence on their evolution.

3.2. Variation and evolution

When considering the evolution of travel guides it is important to underline that, despite the presence of numerous new products on the market, volumes very similar to the first publications are still available and chosen by a large number of consumers. On the other hand, even the most strikingly different new editorial lines display pragmatic and textual features that make the publications easily recognisable as travel guides: the generic integrity is guaranteed by the pragmatic situation of production and use of the books, which in turn reflects the aims of the addresser as well as the needs and expectations of the addressees. The various text types are differently merged and produce radically different solutions, but in any bookshop around the world no one would fail to single out a travel guide among other books, journals or less comprehensive publications. In other words, variation does not compromise the existence of a recognisable genre, though it can be difficult to imagine a single prototype as valid reference for the whole category, unless the external aspect of the text is neglected and attention is focused on its function as well as on the presence of the instructive component which – more or less overtly – 'guides' the reader through the described places.[16]

One of the most evident innovations in the history of travel guides concerns the presence of images: the traditional 19th-century handbooks – and their present-day counterparts – did not include any

---

15 This is the reason why Barthes (1957), discussing the features of the Blue Guides of Spain, affirmed that they actually were the opposite of what they should have been, true means to 'blind' the reader.

16 Kerbrat (2004) talks about 'procedural discourse'. I have described its linguistic manifestations in Italian (Santulli 2007).

type of images apart from maps and town plans. This was the explicit choice of publishers. But things have gradually changed: most publications in the mid-20th century included drawings sketching important monuments and beautiful views; then photographs appeared, first in black and white, then in colour. Initially, photos used to be grouped in a special section, generally printed in glossy paper, but then they started to be inserted into the main text, offering the visual reproduction of objects and places described in the page. In more recent cases, photographic images can reproduce single details, with irregular margins, which strongly attract the attention of the reader; sometimes visual reproductions in axonometric format function as hypertextual 'nodes', surrounded by a series of simultaneously opened 'pop-ups' – chunks of text that give information and illustrate the most important elements or features of the monument.[17]

The massive presence of photos is emphasised by some publishers as a point in favour of their products. Actually, they make a book much heavier and do not improve the knowledge of readers, who could – and possibly should – use their own eyes to see what is shown in photographs. Yet they are nowadays very common in travel guides, and their presence combines with a radical evolution in text design and important changes in language use. From this point of view, travel guides participate in the evolution of textuality that has been particularly evident across different genres and media in the last few decades. The linear presentation of itineraries, characterised by continuity and coherence in textual and verbal presentation, has been replaced by the accumulation of shorter and shorter chunks of text, each referring to a single place, monument or detail. This layout, combined with the more traditional use of typography (font type and dimension, bold, italics, etc), guarantees easier access to single bits of information and apparently leaves more room for personal choice in the construction of itineraries. Actually, the elimination of the linguistic elements that reproduced the movement of the tourist and even oriented his eyes does not imply that the guide has given up its power of imposing a chosen

17 This technique is commonly adopted by the Dorling Kindersley guides, available also in a German version, which on the Italian market have a very similar counterpart, the *Guida Mondadori*.

perspective for the visit. The drastic reduction of verbal description, compensated by the expansion of the visual component, makes the suggestions even more implicit: the reader is confronted with a coloured page and a highly fragmented text, which impose to find in the real world the reproduced objects and virtually exclude any other experience, thus making him 'blind' to any different stimuli.

Though publishers in different countries still offer traditional products which rigorously exclude images, most guides have accepted the innovation, which however has been adopted more drastically by popular publications. As a matter of fact, the range of variation, which is typical of the genre today, largely depends on the evolution and structure of its readership. In the course of the 20th-century, the number of tourists has constantly increased and travelling has progressively attracted all social classes. Obviously, this has caused a differentiation in travelling habits, linked to different available budgets, but also stemming from different cultural interests. Publishers have started to offer more popular series, with less room for historical and artistic information, more and more focused on very few 'symbolic' items: the extreme selection of sites and monuments presents to the reader only what must absolutely be visited; on the other hand, more trivial but attracting information is given, concerning for example food and drinks, traditional events, shopping and entertainment. In this context, the visual component is expanded and the verbal text highly fragmented and reduced to short, juxaposed paragraphs, each describing an artistic item or any more frivolous visiting opportunity. These publications are meant for mass tourism: the ethos of the guide is not explicitly defined in the introduction (which is often omitted), nor is the identity of the addressee. However, it is not difficult to identify the implied reader, a tourist who is most interested in having a pleasant time and does not care much for learning and coming into actual contact with different places and traditions. Popular publications, having lost their own 'voice', often include suggestions of previous readers, presented as personal 'tips'. In this respect, they reproduce the idea of bloggers who post their comments, suggesting, encouraging or discouraging, praising or blaming. The imposing attitude of the guide is skilfully hidden, while the comments of peers (which were obviously selected by the editors) give an impression of free circulation of ideas and judgements.

Parallel to these publications, the sector offers niche products: specialized guides for all types of 'special' travellers. Any cultural or behavioural feature can be selected as distinctive of a group: bike riding or mountaineering, sailing or diving, wine tasting or discovery of ancient traditions, even literary or historical memories. There are guides which accompany the traveller (not to be confused with a 'tourist')[18] to discover the life of a poet or to visit the places described in a novel; guides for sustainable, responsible, ecologic tourism, and so on. When the community of addressees is much smaller and identified by a shared interest, the guide generally tries to establish a closer relationship: it adopts a more informal style or simply opens up a dialogue with its reader, whose presence in the text is marked by personal deixis. The tone of communication often echoes passages of travel diaries by famous authors (as, for example, Dumas, Goethe or Stendhal), who indulged in conversation with their readers and gave them useful suggestions for future trips.

Considering the articulation of the sector, characterised by wide intertextual gaps, more extensive change for the whole genre can be expected, possibly involving (if we accept Halmari and Virtani's perspective) hybridization with other genres. Actually, the influx of travel diaries or of travel reportage (typical of specialised magazines) can be felt in those elite publications which emphasise the importance of personal experience, both of the writer of the guide and of its reader. This pushes them back into the realm of authorship, in the form that predominated before the development of travel guides proper. On the other hand, the concise and detached presentations meant for mass tourism make those guides more similar to travel brochures, with which they share a covert persuasive intention. Nor is the co-existence of different media irrelevant for the development of new generic features: fragmentation, dialogue and interactivity which dominate the web environment influence the style of a large share of contemporary publications.

18 The different denominations are discussed in Antelmi (2010).

Finally, it is worth mentioning the evolution of the overt persuasive component, which has been eliminated from the peritextual parts of the guide. Since their first appearance travel guides have openly declared their aims in their introductive pages, emphasising the experience and reliability of their – though anonymous – authors and even explicitly describing the character of their readers. However, persuasion in the text has always remained totally hidden, argumentative linguistic structures have been avoided and instruction masked with impersonal sequences. In more recent times, the absence of an introductory travelling programme corresponds to an even more detached and fragmented presentation, confirming the idea that speakers (and readers) do not accept to be persuaded, and therefore do not want to be told that there is someone who chose and decided for them. They like to enjoy the advantages of instruction, but do not want to realize that they are being instructed.

4. Conclusion

The case of tourist guides confirms some important points discussed in the first part of this study; namely:

- the historical context plays a crucial role in genre development and evolution, as it gives rise to specific needs, typical of a certain social pattern, which combine with ideological objectives (in this case, with the aim to offer evaluative description of tourist sites and at the same time define the identity of modern tourists);
- a new genre can emerge for such historical reasons (*why*), through the re-interpretation and combination of features typical of other existing genres *(how*), displaying, at the same time, special distinctive traits (most importantly, prominence of the publisher and corresponding absence of an author, presence of an instructive component);

- a genre is a complex reality: it can include different text types with different specific objectives, but this does not imply that it is a 'hybrid' (unless we accept the principle that all genres are);
- a genre can change through the accumulation of new features, and can also display important elements of variation, more or less 'traditional' forms, which do not necessarily imply a loss of its identity; when the intertextual gap increases, however, it is possible for a genre to split, to accept traits typical of another genre or even to merge with it;
- finally, persuasion does influence the evolution of a genre, even when it is not its most distinctive component, yet it cannot be considered the exclusive reason for change.

All these observations could also apply to the description of language and language change, thus confirming the heuristic value of the hypothesis of the existence of similarities between language and genre which was assumed as a starting point for the present discussion.

References

AIFA – Ministero della Salute 2005. Il poster congressuale: come prepararlo, illustrarlo e renderlo efficace. *Bollettino d'Informazione sui Farmaci* 5-6, 237-240.

Amernic, Joel / Craig, Russell / Tourish, Dennis 2007. The Transformational Leader as Pedagogue, Physician, Architect, Commander, and Saint: Five Root Metaphors in Jack Welch's Letters to Stockholders of General Electric. *Human Relations.* 60/12, 1839-1872.

AntConc 3.2.1 Available at <http://www.antlab.sci.waseda.ac.jp/software.htm>.[1]

Antelmi, Donella 2010. Viaggiatori e testi: identità discorsive. *Rivista on-line dell'Associazione Italiana di Studi Semiotici.* IV/6, 35-42.

Arthur Andersen 2000. *Spice up the Story: A Survey of Narrative Reporting in Annual Reports.* London.

Askehave, Inger / Ellerup Nielsen, Anne 2004. Webmediated Genres – A Challenge to Traditional Genre Theory. *Working Paper nr. 6,* Aarhus, Center for Virksomhedskommunication.

Askehave, Inger / Ellerup Nielsen, Anne 2005. What are the Characteristics of Digital Genres? – Genre Theory from a Multi-modal Perspective. *Information Technology & People.* 18/2, 120-141.

Atkinson, Dwight 1999. *Scientific Discourse in Sociohistorical Context. The Philosophical Transactions of the Royal Society of London, 1675-1975.* Mahwah, NJ: Erlbaum.

Baker, Paul 2004. Querying Keywords. Questions of Difference, Frequency, and Sense in Keywords Analysis. *Journal of English Linguistics.* 32/4, 346-359.

Bakhtin, Mikhail M. 1935/1981. *The Dialogic Imagination: Four Essays*, ed. Michael Holquist, transl. Caryl Emerson and Michael Holquist. Austin: University of Texas Press.

Bakhtin, Mikhail 1986. The Problem of Speech Genres. In Bakhtin, Mikhail, *Speech Genres and Other Late Essays.* Caryl Emerson and Michael Holquist (eds). Transl. Vern W. McGee. Austin: University of Texas Press, 60-102.

1 Unless otherwise indicated, all websites were last accessed in January 2012.

Barbiers, Sjef 2002. An Introduction to Modality and its Interaction with the Verbal System. In Barbiers, Sjef / Beukema, Frits / van der Wurff, William (eds) *Modality and its Interaction with the Verbal System*. Amsterdam: John Benjamins, 1-17.

Barthes, Roland 1957. *Mythologies*. Paris: Seuil.

Barthes, Roland 1970/1974 *S/Z*, transl. Richard Miller, New York: Hill and Wang (*S/Z*, Paris, Éditions du Seuil).

Barton, David / Hall, Nigel (eds) 1999. *Letter Writing as a Social Practice*. Amsterdam/Philadelphia: John Benjamins.

Barton, Ellen L. 1993. Evidentials, Argumentation and Epistemological Stance. *College English*. 55/7, 745-769.

Battalio, John T. 1998. *The Rhetoric of Science in the Evolution of American Ornithological Discourse*. Stanford, CA: Ablex.

Bawarshi, Anis S. / Reiff, Mary Jo 2010. *Genre: An Introduction to History, Theory, Research, and Pedagogy*. Indiana: Parlor Press and the WAC Clearinghouse.

Bazerman, Charles 1988. *Shaping Written Knowledge. The Genre and Activity of the Experimental Article in Science*. Madison, WI: The University of Wisconsin Press.

Bazerman, Charles 1994. Systems of Genres and the Enactment of Social Intentions. In Freedman, Aviva / Medway, Peter (eds), 79-101.

Bazerman, Charles 2008. Theories of the Middle Range in Historical Studies of Writing Practice. *Written Communication*. 25/3, 298-318.

Bell, Alan 1991. *The Language of News Media*. Oxford: Blackwell.

Bellier, Irène 2005. Anthropology of Institutions and Discourse Analysis. Looking into Interdisciplinarity. In Wodak, Ruth / Chilton, Paul (eds) *A New Agenda in (Critical) Discourse Analysis. Theory, Methodology and Interdisciplinarity*. Amsterdam: John Benjamins, 243-267.

Benoit, William L. 1995. *Accounts, Excuses and Apologies. A Theory of Image Restoration Strategies*, Albany (NY): State University of New York Press.

Benoit, William L. 2000. Another Visit to the Theory of Image Restoration Strategies. *Communication Quarterly*. 48/1, 40-44.

Berge, Kjell Lars 1994. The Diachrony of Textual Norms; Or, Why Do Genres Change? A Presentation of a Theory Illustrated by Its Relevance for the Teaching of Writing. *Changing English*. 1/1, 217-228.

Berkenkotter, Carol 2008a. Genre evolution? The case for a diachronic perspective. In Bhatia, Vijay K. / Flowerdew, John / Jones, Rodney H. (eds) *Advances in Discourse Studies*. London / New York: Routledge, 178-191.

Berkenkotter, Carol 2008b. *Patient Tales: Case Histories and the Uses of Narrative in Psychiatry.* Columbia: University of South Carolina Press.

Berkenkotter, Carol / Huckin, Thomas N. 1993. Rethinking Genre from a Sociocognitive Perspective. *Written Communication.* 10/4, 475-509.

Berkenkotter, Carol / Huckin, Thomas N. 1995. *Genre Knowledge in Disciplinary Communication: Cognition, Culture, Power.* Hillsdale, NJ: Lawrence Erlbaum Associates.

Bernal, Sebastiá / Chillón, Lluís Albert 1985. *Periodismo informativo de creación*. Barcelona: Mitre.

Bernardini, Piero 2000. *L'arbitrato commerciale internazionale*. Milano: Giuffrè.

Bhatia, Vijay K. 1993. *Analysing Genre: Language Use in Professional Settings*. London / New York: Longman.

Bhatia, Vijay K. 1994. Generic Integrity in Professional Discourse. In Gunnarsson, Britt-Louise / Linell, Per / Nordberg, Bengt (eds) *Text and Talk in Professional Contexts*. Uppsala: ASLA, 61-76.

Bhatia, Vijay K. 1995. Genre-mixing in Professional Communication: The Case of 'Private Intentions' v. 'Socially Recognized Purposes'. In Bruthiaux, Paul / Boswood, Tim / Du-Babcock, Bertha (eds) *Explorations in English for Professional Communication*. Hong Kong: Department of English, City University of Hong Kong, 1-19.

Bhatia, Vijay K. 1997a. Genre Mixing in Academic Introductions. *English for Specific Purposes.* 16/3, 181-195.

Bhatia, Vijay K. 1997b. The Power and Politics of Genre. *World Englishes*. 16/3, 359-371.

Bhatia, Vijay K. 2004. *Worlds of Written Discourse. A Genre-based View*. London: Continuum.

Bhatia, Vijay K. 2007. Interdiscursivity in Critical Genre Analysis. *Proceedings of the 4th International Symposium on Genre Studies*. Available at: <http://www3.unisul.br/paginas/ensino/pos/linguagem/cd/English/36i.pdf> .

Bhatia, Vijay K. 2008a. Towards Critical Genre Analysis. In Bhatia, Vijay K. / Flowerdew, John / Jones, Rodney H. (eds) *Advances in Discourse Studies.* London / New York: Routledge, 167-177.

Bhatia, Vijay K. 2008b. Genre Analysis, ESP and Professional Practice. *English for Specific Purposes.* 27: 161-174.

Bhatia, Vijay K. 2010. Interdiscursivity in professional communication. *Discourse & Communication.* 21/ 1: 32-50.

Bhatia, Vijay K. / Candlin, Christopher N. / Gotti, Maurizio (eds) 2004. *Legal Discourse in Multilingual and Multicultural Contexts. Arbitration Texts in Europe*. Bern: Peter Lang.

Bhatia, Vijay K. / Candlin, Christopher N. / Gotti, Maurizio (eds) 2010. *The Discourses of Dispute Resolution*. Bern: Peter Lang.

Biber, Douglas / Connor, Ulla / Upton, Thomas A. 2007. *Discourse on the Move*. Amsterdam: John Benjamins.

Block, Steven M. 1996. Do's and Don'ts of Poster Presentation. *Biophysical Journal*. 71, 3527-3529.

Bolter, Jay D. / Grusin, Richard 1999. *Remediation. Understanding New Media*. Cambridge MA.: The MIT Press.

Bondi, Marina 1997. The Rise of Abstracts. Development of the Genre in the Discourse of Economics. *TEXTUS. Special Issue: Linguistic Change in English: A Socio-Cultural Overview*. Barisone Ermanno / Hughes, Geoffrey (eds). 10/2, 395-418.

Bondi, Marina 2001. Small Corpora and Language Variation. Reflexivity across Genres. In Ghadessy, Mohsen / Henry, Alex / Roseberry, Robert L. (eds) *Small Corpus Studies and ELT*. Amsterdam: John Benjamins, 135-174.

Bondi, Marina 2004. The Discourse Function of Academic Connectors in Abstracts. In Stenström, Anna-Brita / Aijmer,Karin (eds) *Discourse Patterns in Spoken and Written Corpora*. Amsterdam: John Benjamins, 139-156.

Bondi, Marina 2010. Metadiscursive Practices in Introductions. Phraseology and Semantic Sequences across Genres. *NJES (Nordic Journal of English Studies)*. 9/2, 99-123.

Boorstin, Daniel 1986. *Les Découvreurs*. Paris: Seghers.

Bournois, Frank / Point, Sébastien 2006. A Letter from the President: Seduction, Charm and Obfuscation in French CEO Letters. *Journal of Business Strategy*. 27/6, 46-55.

Brilli, Attilio 1995. *Quando viaggiare era un'arte*. Bologna: il Mulino.

Brown, Chris / Ainley, Kirsten 2005. *Understanding International Relations*. Basingstoke: Palgrave.

Brown, Halina Szejnwald / de Jong, Martin / Lessidrenska, Teodorina 2009. The Rise of the Global Reporting Initiative: A Case of Institutional Entrepreneurship. *Environmental Politics*. 18/2, 182-200.

Burgess, Sally / Fagan, Anna. 2004. Genre Anaysis and the Novice Researcher: The Book Review and the Poster Presentation. In Brito, Manuel / Oliva, Juan Ignacio (eds) *Traditions and Innovations: Commemorating Forty Years of English Studies at ULL (1963-2003)*. La Laguna: RCEI, 61-77.

Busch-Lauer, Ines-A. 2003. Perspective in Medical Correspondence: English and German Letters-to-the-Editor. In Ensink, Titus / Sauer, Christoph (eds) *Framing and Perspectivising in Discourse*. Amsterdam: John Benjamins, 191-214.

Bybee, Joan L. / Fleischman, Susan (eds) 1995. *Modality in Grammar and Discourse*. Amsterdam: John Benjamins.

Caliendo, Giuditta 2004. Modality and Communicative Interaction in EU Law. In Gotti, Maurizio / Candlin, Christopher N. (eds) *Intercultural Aspects of Specialized Communication*. Bern: Peter Lang, 241-259.

Caliendo, Giuditta / Di Martino, Gabriella / Venuti, Marco 2005. Language and Discourse Features of EU Secondary Legislation. In Cortese, Giuseppina / Duszak, Anna (eds) *Identity, Community, Discourse: English in Intercultural Settings*. Bern: Peter Lang, 381-404.

.Calsamiglia, Helena / López Ferrero, Carmen 2003. Role and Position of Scientific Voices: Reported Speeches in the Media. *Discourse Studies*. 5/2, 147-173.

Calsamiglia, Helena / van Dijk, Teun A. 2004. Popularization Discourse and Knowledge about the Genome. *Discourse and Society*. 15/4, 370-389.

Campbell, Karlyn Kohrs / Hall Jamieson, Kathleen 1990. *Deeds Done in Words: Presidential Rhetoric and the Genres of Governance*. Chicago: University of Chicago Press.

Candlin, Christopher N. / Maley, Yon 1997. Intertextuality and Interdiscursivity in the Discourse of Alternative Dispute Resolution. In Gunnarsson, Britt-Louise / Linnel, Per / Nordberg, Bengt (eds) *The Construction of Professional Discourse*. London: Longman, 201-222.

Catenaccio, Paola 2010. Genre Repurposing and New Communicative Exigencies: The Case of the CEO's Letter in CSR Reports. *RILA*, XLII/3, 185-204.

Catenaccio, Paola 2011a. Social and Environmental Reports: A Short-term Diachronic Perspective on an Emerging Genre. In Garzone, Giuliana / Gotti, Maurizio (eds) *Discourse, Communication and the Enterprise. Genres and Trends*. Bern: Peter Lang, 169-192.

Catenaccio, Paola 2011b. Green Advertising and Corporate CSR Communication: Hybrid Discourses, Mutant Genres. In Sarangi/Polese/Caliendo (eds), 353-372.

Catenaccio, Paola / Degano, Chiara 2011. Corporate Social Responsibility as a Key to the Representation of Corporate Identity: The Case of Novartis. In Evangelisti Allori, Paola. / Bhatia, Vijay K. (eds) *Identity in the Professions. Legal, Corporate and Institutional Citizenship*. Bern: Peter Lang, 79-102.

Charaudeau, Patrick 1997. *Le discours d'information médiatique: La construction du miroir social*. Paris: Nathan.

Chesterman, Simon 2004. *You, the People: The United Nations, Transitional Administration, and State-Building.* Oxford: Oxford University Press.

Chouliaraki, Lilie / Fairclough, Norman 1999. *Discourse in Late Modernity. Rethinking Critical Discourse Analysis*. Edinburgh: Edinburgh University Press.

Coates, Jennifer 1983. *The Semantics of the Modal Auxiliaries*. London / Canberra: Croom Helm.

Conaway, Roger N. / Wardrope, William J. 2010. Do their Words Really Matter? Thematic Analysis of U.S. and Latin American CEO Letters. *Journal of Business Communication.* 47/2, 141-168.

Conger, Cristen 2009. *How Did a Harvard Commencement Speech Help Save Europe after World War II?* Article posted on 12/02/09 and available at <www.howstuffworks.com>.

Cook, Guy 2001. *The Discourse of Advertising*. London: Routledge ([1]1992).

Coseriu, Eugene 1974. *Synchronie, Diachronie und Geschichte: das Problem des Sprachwandels*. München: Fink.

Coupland, Christine 2005. Corporate Social Responsibility as Argument on the Web. *Journal of Business Ethics.* 64/2, 355-366.

Coupland, Christine 2006. Corporate Social and Environmental Responsibility in Web-based Reports: Currency in the Banking Sector? *Critical Perspectives in Accounting*. 17, 865-881.

Culler, Jonathan 1988. *Framing the Sign: Criticism and its Institutions.* London: University of Oklahoma Press.

Cutts, Martin 1999. *The Plain English Guide.* Oxford: Oxford University Press.

Cutts, Martin 2000. *Lucid Law, 2nd edition.* Plain Language Commission. <http://www.clearest.co.uk/files/LucidLawA-T.pdf>.

D'Avanzo, Stefania 2010. *EU Political and Legal Discourse on Immigration and Asylum.* Doctoral Dissertation in *Lingua Inglese per Scopi Speciali* (English for Special Purposes), University of Naples 'poli Federico II' (<www.fedoa.unina.it>).

D'Avanzo, Stefania 2011. European Summaries of Directives on Asylum: Hybridization in Institutional Discourse. In Sarangi/Polese/Caliendo (eds), 87-97.

Dahl, Trine 2004. Textual Metadiscourse in Research Articles: a Matter of National Culture or of Academic Discipline? *Journal of Pragmatics*. 36, 1807-1825.

Dahl, Trine 2009. Author Identity in Economics and Linguistics Abstracts. In Suomela-Salmi, Eija / Dervin Fred (eds) *Cross-lingustic and Cross-*

cultural Perspectives on Academic Discourse. Amsterdam: John Benjamins, 123-134.

De Bens, Els / Raeymaeckers, Karin 2010. *De Pers in België: het verhaal van de Belgische Dagbladpers: Gisteren, Vandaag en Morgen*. Tielt: LannooCampus.

De Castro, Paola. 2009. *Librarians of Babel. A Toolkit for Effective Communication*. Oxford: Chandos Publishing.

De Fina, Anna 2009. Narratives in Interview – The Case of Accounts, *Narrative Inquiry*. 19/2, 233-258.

De Groot, Elizabeth 2008. *English Annual Reports in Europe: A Study on the Identification and Reception of Genre Characteristics in Multimodal Annual Reports Originating in the Netherlands and in the United Kingdom*. Utrecht: LOT.

Degano, Chiara 2008. The Construction of Cultural Identity in Banks' Reports on Corporate Social Responsibility. In Di Martino, Gabriella / Polese, Vanda / Solly, Martin (eds) *Identity in English Domain-Specific Discourse*. Naples: Edizioni Scientifiche Italiane (ESI), 21-41.

Degano, Chiara 2010. Indicators of Argumentation in Arbitration Awards: A Diachronic Perspective. In Bhatia/Candlin/Gotti (eds) 2010, 189-205.

Demarteau, Nadia / Moeremans, Karen / Annemans, Lieven 2007. Critical Appraisal of Scientific Posters Comparing Anemia Treatments for Cancer Patients: Applying ISPOR Task Force Guidelines on Methodological Quality of Retrospective Studies. *Critical Reviews in Oncology/Hematology*. 63, 91-99.

Department of Conference Services 1984. *A Guide to Writing for the United Nations*. New York: United Nations.

Derrida, Jacques 1972. *La dissémination*. Paris : Éditions du Seuil.

Devitt, Amy 1991. Intertextuality in Tax Accounting: Generic, Referential, and Functional. In Bazerman, Charles / Paradis, James (eds) *Textual Dynamics of the Professions*. Madison (WI): University of Wisconsin Press, 306-335.

Devitt, Amy J. 2004. *Writing Genres*. Carbondale: Southern Illinois University Press.

Di Blasio, Norina W. 2004. La Comunicazione Congressuale. Il Poster. In De Castro, Paola / Guida, Silvana / Sagone, Bianca Maria (eds) *Diciamolo Chiaramente*. Roma: Il Pensiero Scientifico Editore, 229-239.

Dieltjens, Sylvain / Heynderickx, Priscilla 2005. Veel geblaat, weinig wol. Vacatures in het vizier. *Ad Rem*. 17/5, 6-9.

DiStaso, Marcia 2008. Lessons Learned from 13 Financial Communicators. *The Measurement Standard. The Newsletter of Public Relations and*

Social Media Measurement from KDPaine and Partners. Online at <http://kdpaine.blogs.com/themeasurementstandard/>.

Docherty, Michael / Smith, Richard 1999. The Case for Structuring the Discussion of Scientific Papers. *British Medical Journal*. 318, 1221-1222.

Dos Santos, Mauro. B. 1996. The Textual Organization of Research Paper Abstracts in Applied Linguistics. *Text*. 16/4, 481-499.

Durnez, Gaston 1998. De Standaard. In De Schrijver, Reginald *et al*. (eds) *Nieuwe Encyclopedie van de Vlaamse Beweging*. Lannoo: Tielt.

Duwe, David 2005. *The Miami Herald and the Miller Effect Literary Journalism in the 1980s*. University of Florida. Available at: <http://etd.fcla.edu/UF/UFE0012301/stantond.pdf>.

Ellerbee, Susan 2009. An Artistic View of Posters. *Trends in Neuroscience*. 9/2, 109-110.

Ellerup Nielsen, Anne / Thomsen, Christa 2007. Reporting CSR – What and How to Say It? *Corporate Communications: An International Journal*. 12/1, 25-40.

Erren, Thomas C. / Bourne, Philip E. 2007. Ten Simple Rules for a Good Poster Presentation. *PLoS Computational Biology*. 3/5, 777-778.

Fairclough, Norman 1989. *Language and Power*. London / New York: Longman.

Fairclough, Norman 1993. Critical Discourse Analysis and the Marketization of Public Discourse: The Universities. *Discourse & Society*. 4/2, 133-168.

Fairclough, Norman 1995. *Critical Discourse Analysis*. Harlow: Longman.

Fairclough, Norman 2003. *Analysing Discourse: Textual Analysis for Social Research*. London/New York: Routledge.

Fairclough, Norman 2005. Governance, Partnership and Participation: Cooperation and Conflict. Unpublished paper given at the conference of the International Association for Dialogue Analysis, Bucharest. Available at <http://www.ling.lancs.ac.uk/profiles/263>.

Fairclough, Norman 2007. Introduction. In Cortese, Giuseppina / Fairclough, Norman / Ardizzone, Patrizia (eds) *Discourse and Contemporary Social Change*. Bern: Peter Lang, 9-14.

Fairclough, Norman / Wodak, Ruth, 1997. Critical Discourse Analysis. In van Dijk, Teun A. (ed.) *Discourse as Social Interaction*. London: Sage, 258-284.

Fanelli, Fanelli / Misangyi, Vilmos F. / Tosi, Henry L. 2009. In Charisma We Trust: The Effects of CEO Charismatic Visions on Securities Analysts. *Organization Science*. 20, 1011-1033.

Flowerdew, John / Wan, Alina 2010. The Linguistic and the Contextual in Applied Genre Analysis: The Case of the Company Audit Report. *English for Specific Purposes.* 29, 78-93.

Flowerdew, Lynne 2005. An integration of corpus-based and genre-based approaches to text analysis in EAP/ESP: countering criticisms against corpus-based methodologies. *English for Specific Purposes.* 24/1, 321-332.

Freddi, Maria 2005. Arguing Linguistics: Corpus Investigation of One Functional Variety of Academic Discourse. *Journal of English for Academic Purposes.* 4/1, 5-26.

Freedman, Aviva / Medway, Peter (eds) 1994. *Genre and the New Rhetoric.* London: Francis & Taylor.

Garzone, Giuliana 1997. Osservazioni sul valore pedagogico dell'ipertesto. In Schena, Leo / Garzone, Giuliana (eds). *Il computer al servizio degli insegnamenti linguistici, Milano 28.4.1995-3.5.1996.* Milan: Centro Linguistico Università Bocconi, 21-39.

Garzone, Giuliana 2002. Describing E-commerce Communication. Which Models and Categories for Text Analysis? *TEXTUS. Special Issue: English in Academic and Professional Settings: Techniques of Description / Pedagogical Application.* Evangelisti, Paola / Ventola, Eija (eds). XIV/2, 279-296.

Garzone, Giuliana 2001. Deontic Modality and Performativity in English Legal Texts. In Gotti, Maurizio / Dossena, Marina (eds) *Modality in Specialized Texts.* Bern: Peter Lang, 153-173.

Garzone, Giuliana 2004a. Arbitration Rules Across Legal Cultures: An Intercultural Approach. In Bhatia/Candlin/Gotti (eds) 2004, 177-220.

Garzone, Giuliana 2004b. Annual Company Reports and CEOs' Letters: Discoursal Features and Cultural Markedness. In Candlin, Christopher N. / Gotti, Maurizio (eds) *Intercultural Aspects of Specialized Communication.* Bern: Peter Lang, 311-342.

Garzone, Giuliana 2005a. Letters to Shareholders and Chairman's Statements: Textual Variability and Generic Integrity. In Gillaerts/Gotti (eds), 179-204.

Garzone, Giuliana 2005b. Annual Company Reports and CEOs' Letters: Discoursal Features and Cultural Markedness. In Candlin Christopher N. / Gotti, Maurizio (eds) *Intercultural Aspects of Specialized Communication.* Bern: Peter Lang, 311-342.

Garzone, Giuliana 2006. *Perspectives on ESP and Popularization.* Milano: CUEM.

Garzone, Giuliana 2007. Genres, Multimodality and the World Wide Web: Theoretical Issues. In Garzone, Giuliana / Poncini, Gina / Catenaccio, Paola (eds) *Multimodality in Corporate Communication. Web Genres and Discursive Identity*. Milan: FrancoAngeli.

Garzone, Giuliana 2008. Gli enunciati performativi nel testo giuridico inglese: la prospettiva linguistica. In Garzone, Giuliana / Santulli, Francesca (a cura di) *Il linguaggio giuridico. Prospettive interdisciplinari*. Milano: Giuffré Editore, 57-87.

Garzone, Giuliana 2009. Multimodal Analysis. In Bargiela-Chiappini, Francesca (ed.) *The Handbook of Business Discourse*. Edinburgh: Edinburgh University Press, 154-165.

Garzone, Giuliana 2010. Arbitration and Mediation on the Web. In Bhatia/ Candlin/Gotti (eds), 23-46.

Gaynesford de, Robert M. 2006. *The Meaning of the First Person Term*. Oxford: Clarendon Press.

Genette, Gerard 1972. *Figures III*. Paris: Seuil.

Gergen, Kenneth J. / Graumann, Charles F. (eds). 2006. *Psychological Discourse in Historical Perspective*. New York: Cambridge University Press.

Giannoni, Davide Simone 2006. Evidence of Generic Tension in Academic Book Acknowledgements. In Bhatia, Vijay K. / Gotti, Maurizio (eds) *Explorations in Specialized Genres*. Bern: Peter Lang, 21-42.

Giannoni, Davide Simone 2008a. Medical Writing at the Periphery: The Case of Italian Journal Editorials. *Journal of English for Academic Purposes*. 7/2, 97-107.

Giannoni, Davide Simone 2008b. Disciplinary and Linguistic Identities in the Journal Editorial Genre. In Di Martino, Gabriella / Polese, Vanda / Solly, Martin (eds) *Identity and Culture in English Domain-specific Discourse*. Naples: Edizioni Scientifiche Italiane, 325-348. Preprint available at <http://wwwdata.unibg.it/dati/persone/991/2701.pdf>.

Giglioni, Cinzia 2010. Apologetic Discourse in CEO and Chairman's Statements: The FT30 Index. *IR Top. Journal of Investor Relations*. IX/4, 29-34.

Gillaerts, Paul / Gotti, Maurizio (eds) 2005. *Genre Variation in Business Letters*. Bern: Peter Lang.

Gillaerts Paul / van de Velde Freek 2010. Interactional Metadiscourse in Research Article Abstracts. *Journal of English for Academic Purposes*. 9/2, 128-139.

Giltrow, Janet / Stein, Dieter 2009. *Genres in the Internet*. Amsterdam: John Benjamins.

Goddard, Angela 2002. *The Language of Advertising: Written Texts*. London: Routledge.

Goffman, Erving 1971. Remedial Interchanges. In *Relations in Public: Micro-studies of the Public Order*. New York: Harper & Row, 95-187.

Goffman, Erwin 1981. *Forms of Talk*. Oxford: Blackwell.

Golebiowski, Zosia 2009. Prominent Messages in Education and Applied Linguistic Abstracts: How do Authors Appeal to Their Prospective Readers? *Journal of Pragmatics*. 41/4, 753-769.

Gotti, Maurizio 2003. *Specialized Discourse. Linguistic Features and Changing Conventions*. Bern: Peter Lang.

Gotti, Maurizio 2005. *Investigating Specialized Discourse.* Bern: Peter Lang.

Gotti, Maurizio (ed.) 2009. *Commonality and Individuality in Academic Discourse*. Bern: Peter Lang.

Goźdź-Roszkowski, Stanisław 2010. *Responsibility* and *Welfare*: Keywords and Semantic Categories in Legal Academic Journals. In Giannoni, Davide Simone / Frade, Celina (eds) *Researching Language and the Law. Textual Features and Translation Issues*. Bern: Peter Lang, 71-87.

Graham, Philip 2001. Contradictions and Institutional Convergences: Genre as Method. *Journal of Future Studies*. 5/4, 1-30.

Grant Thornton 2011. *Corporate Social Responsibility: The Power of Perception.* <http://www.grantthornton.co.nz/Assets/documents/pubSeminars/IBR-2011- CSR-Report.pdf>.

Grant, Ruth W. / Keohane, Robert O. 2005. Accountability and Abuses of Power in World Politics. *American Political Science Review.* 99/1, 29-43.

GRI (Global Reporting Initiative) 2006. *Sustainability Reporting Guidelines G3*. Amsterdam: GRI.

Gross, Alan G. / Harmon, Joseph E. / Reidy, Michael 2002. *Communicating Science: The Scientific Article from the Seventeenth Century to the Present*. New York: Oxford University Press.

Gunningham, Neil 2009. Shaping Corporate Environmental Performance: A Review. *Environmental Policy and Governance*. 19, 215-231.

Gunter, Barrie 2003. *News and the Net*. Mahwah, New Jersey: Lawrence Erlbaum Associates.

Halliday, Michael A.K. / Martin, Jim R. (eds) 1993. *Reading Science: Literacy as Discursive Power*. London: Falmer.

Halliday, Michael A.K. 1994. *An Introduction to Functional Grammar* ([1]1985). London: Arnold.

Halmari, Helena / Virtanen, Tuija (eds) 2005a. *Persuasion Across Genres*. Amsterdam / Philadelphia: John Benjamins.

Halmari, Helena / Virtanen, Tuija 2005b. Towards Understanding Modern Persuasion. In Halmari /Virtanen (eds), 229-244.

Hansen, Anders / Machin, David 2008. Visually Branding the Environment: Climate Change as a Marketing Opportunity. In *Discourse Studies*. 10/6, 777-794.

Harrell, Jackson / Ware, B. L. / Linkugel, Wil A. 1975. Failure of Apologia in American Politics: Nixon on Watergate. *Speech Monographs*. 42, 245-261.

Harris, Sandra / Grainger, Karen / Mullany, Louise. 2006. The Pragmatics of Political Apologies. *Discourse & Society*. 17, 715-747.

Harrison, Teresa M. / Barthel, Brea 2009. Wielding New Media in Web 2.0: Exploring the History of Engagement with the Collaborative Construction of Media Products. *New Media & Society*. 11/1-2, 155-178.

Hearit, Keith M. 2006. *Crisis Management by Apology. Corporate Response to Allegations of Wrongdoing*. Mahwah (NJ): Lawrence Erlbaum Associates Publishers.

Heynderickx, Priscilla / Dieltjens, Sylvain 2000. De nieuwjaarsboodschap als pictogram. *Ad Rem*. 14/4, 1-5.

Heynderickx, Priscilla / Dieltjens, Sylvain 2005. De Personeelsadvertentie onder de Loep genomen. In Mottard, André (ed.) *Retoriek en Praktijk van het moedertaalonderwijs: een Selectie uit de 18de Conferentie Het schoolvak Nederlands*. Gent: Academia Press, 29-45.

Hilgartner, Stephen 1990. The Dominant View of Popularization: Conceptual Problems, Political Uses. *Social Studies of Science*. 20/3, 519-539.

Hjelmslev, Louis 1963. *Sproget. En introduction*. København: Berlingkse Forlag.

Hoffman, Donna L. / Novak, Tom P. 1996. Marketing in the Hypermedia Computer-Mediated Environment, Conceptual Foundations. *Journal of Marketing*. 60/3, 50-68.

Horton, Richard 1995. The Rhetoric of Research. *British Medical Review*. 310, 985-987.

Hoye, Leo F. 1997. *Adverbs and Modality in English*. London: Longman.

Hoye, Leo F. 2009. Modality in Discourse: The Pragmatics of Epistemic Modality. In Tsangalidis, Anastasios / Facchinetti, Roberta (eds) *Studies on English Modality. In Honour of Frank Palmer*. Bern: Peter Lang, 99-131.

Huang, Jun Ru 2010. 走出科普教化思維 (Moving Away from the Educational Mindset in Popular Science). <http://case.ntu.edu.tw/SciMonthly40/sm40blog/?p=484>.

Hunston, Susan 2002. *Corpora in Applied Linguistics*. Cambridge: Cambridge University Press.

Hunston, Susan 2008. Starting with the Small Words: Patterns, Lexis and Semantic Sequences. *International Journal of Corpus Linguistics.* 13/3, 271-295.

Hunston, Susan / Francis, Gill 2000. *Pattern Grammar. A Corpus-Driven Approach to the Lexical Grammar of English.* Amsterdam / New York: John Benjamins.

Hunston, Susan / Thompson, Geoff (eds) 2000. *Evaluation in Text. Authorial Stance and the Construction of Discourse.* Oxford: Oxford University Press.

Hunter, Martin / Marriott, Arthur / Veder, V. V. 1995. *The Internationalisation of Arbitration.* London: Graham and Trotman.

Hüttner, Julia / Smit, Ute / Mehlmauer-Larcher, Barbara 2009. ESP Teacher Education at the Interface of Theory and Practice: Introducing a Model of Mediated Corpus-based Genre Analysis. *System.* 37/1, 99-109.

Hyland, Ken 1998. Exploring Corporate Rhetoric: Metadiscourse in the CEO's Letter. *Journal of Business Communication.* 35/2, 224-225.

Hyland, Ken 2000. *Disciplinary Discourses.* London: Longman.

Hyland, Ken 2004. *Genre and Second Language Writing.* Michigan: The University of Michigan Press.

Hyland, Ken 2005a. Stance and Engagement: A Model of Interaction in Academic Discourse. *Discourse Studies.* 7/2, 173-192.

Hyland, Ken 2005b. *Metadiscourse. Exploring Interaction in Writing.* London: Continuum.

Hymes, Dell 1974. *Foundations in Sociolinguistics. An Ethnographic Approach.* Philadelphia: University of Pennsylvania Press.

Hyon, Sunny 1996. Genres in Three Traditions: Implications for ESL. *TESOL Quarterly.* 30, 693-722.

Igalens, Jacques 2006. *L'analyse du discourse de la responsibilité sociale de l'enterprise à travers les rapport annuels de développement durable d'enterpris françaises du CAC 40.* Cahier de recherche no. 2006-173. Toulouse: Centre de Recherce en Gestion.

International Committee of Medical Journal Editors (ICMJE) 2009. Uniform requirements for the manuscripts submitted to biomedical journals. *British Journal of Medicine (BJM).* Available at <http://www.icmje.org>.

Jaime Sisó, Mercedes 2009. Titles or Headlines? Anticipating Conclusions in Biomedical Research Article Titles as a Persuasive Journalistic Strategy to Attract Busy Readers. *Miscelánea. A Journal of English and American Studies.* 39, 29-54.

Jameson, Daphne A. 2000. Telling the Investment Story: A Narrative Analysis of Shareholder Reports. *Journal of Business Communication.* 37/1, 7-38.

Jasinski, James 2001. *Sourcebook on Rhetoric. Key Concepts in Contemporary Rhetorical Studies*. London: Sage.

Jáuregui, Pablo 2000. Las células madre neuronales de adulto son pluripotenciales. 3 June 2000. Suplemento *Salud, El Mundo.*

Jauss, Hans-Robert 1977. *Alterität und Modernität der mittelalterlichen Literatur*. München: Fink Verlag.

Jiang, Zhenxian 1990. 台灣七零年代科學普及與科學權威之意識型態：以《科學月刊》為分析個案 (*Science Popularization and the Authority of Science in Taiwan in the 1970s: A Case Study of Science Monthly*). Unpublished MA dissertation, National Tsing-Hua University.

Johnstone, Ian 2003. The Role of the UN Secretary- General: The Power of Persuasion Based on Law. *Global Governance.* 9/3, 441-458.

Jucker, Andreas H. 2006. "but 'tis believed that […]": Speech and Thought Presentation in Early English Newspapers. In Brownlees, Nicholas (ed.) *News Discourse in Early Modern Britain.* Bern: Peter Lang, 105-125.

Kaldor, Mary 1999. *New and Old Wars. Organized Violence in a Global Era.* Cambridge: Polity Press.

Kaplan, Andreas / Haenlein, Michael 2010. Users of the World, Unite! The Challenges and Opportunities of Social Media. *Business Horizons.* 53/1, 59-68.

Kaplan, Robert / Cantor, Selena / Hagstrom, Cynthia / Kamhi-Stein, Lia / Shiotani, Yumiko / Zimmerman, Cheryl 1994. On Abstract Writing. *Text*. 14/3, 401-426.

Kärnä, Jari / Juslin, Heikki / Ahonen, Virpi / Hansen, Eric 2001. Green Advertising, Greenwash or a True Reflection of Marketing Strategies? *Greener Management International: The Journal of Corporate Environmental Strategy and Practice (GMI)*. 33, 59-70.

Keegan, David A. / Bannister, Susan L. 2003. Effect of Colour Coordination of Attire with Poster Presentation on Poster Popularity. *Canadian Medical Association*. 169/12, 1291-1292.

Keller, Eric 1979. Gambits: Conversational Strategy Signals. *Journal of Pragmatics.* 3, 219-238.

Kerbrat, Chaterine 2004. Suivez la guide! Les modalités de l'invitation au voyage dans les guides touristiques: l'exemple de l'île d'Aphrodite. In Baider, Fabienne / Burger, Marcel / Goustos, Dionys (eds) *La communication touristique*. Paris: L'Harmattan, 133-150.

Koesten, Joy / Rowland, Robert C. 2004. The Rhetoric of Atonement. *Communication Studies*. 55, 68-87.

Kolk, Ans 2008. Sustainability, Accountability and Corporate Governance: Exploring Multinationals Reporting Practices. *Business Strategy and the Environment.* 17/1, 1-15.

Kolk, Ans 2010. Trajectories of Sustainability Reporting by MNCs. *Journal of World Business*. 45/4, 367-374.

Kong, Kenneth 2004. Marked Themes and Thematic Patterns in Abstracts, Advertisements and Administrative Documents. *Word.* 55/3, 343-362.

Korswagen, Cees 1985. *Doeltreffend werven, soliciteren en selecteren: een praktische handleiding voor selecteur en sollicitant.* Van Loghum Slaterus: Deventer.

KPMG Global Sustainability Services 2011. *KPMG International Survey of Corporate Responsibility Reporting 2011*. Amsterdam: KPMG International.

Kress, Gunther 1987. Genre in a Social Theory of Language: A Reply to John Dixon. In Reid, Ian (ed.) *The Place of Genre in Learning: Current Debates.* Geelong: Deakin University, 35-45.

Kress, Gunther 2010. *Multimodality. A Social Semiotic Approach to Contemporary Communication.* London: Routledge.

Kress, Gunther / Hodge, Robert 1993. *Language as Ideology*. London: Routledge & Kegan Paul ([1]1979).

Kress, Gunther / van Leeuwen, Theo 2001. *Multimodal Discourse*. London: Arnold.

Kress, Gunther / van Leeuwen, Theo 2004. *Reading Images*. London: Routledge.

Kreuz, Roger J. / Graesser, Arthur C. 1993. The Assumptions behind Questions in Letters to Advice Columnists. *Text.* 13/1, 65-89.

Landow, George P. 1992. *Hypertext: The Convergence of Contemporary Critical Theory and Technology*. Baltimore: Johns Hopkins University Press.

Lee, Tom 1994. The Changing Form of the Corporate Annual Report. *Accounting Historians Journal.* 21/1, 215-232.

Leech, Geoffrey 1966. *English in Advertising*. London: Longman.

Leech, Geoffrey / Short, Mick 2007. *Style in Fiction: A Linguistic Introduction to English Fictional Prose. Second Edition.* Harlow: Pearson Education Limited ([1]1981).

Lemke, Jay L. 2006. Towards Critical Multimedia Literacy: Technology, Research, and Politics. In Reinking, David / Labbo, Linda / McKenna, Michael / Kiefer, Ronald (eds) *International Handbook of Literacy & Technology, v2.0.* Hillsdale NJ: Lawrence Erlbaum Associates, 312-332.

Lew, Robert 1997. Towards a Taxonomy of Linguistic Jokes. *Studia Anglica Posnaniensia*. 31, 132-152.

Lin, Xiaoxin 2010. 科學月刊籌備經過 (The Preparation Process of Science Monthly). Online at <http://case.ntu.edu.tw/SciMonthly40/sm40blog/?p=74#more-74>.

Livesey, Sharon M. 2001. Eco-Identity as Discursive Struggle: Royal Dutch/Shell, Brent Spar, and Nigeria. *Journal of Business Communication.* 38, 58-91.

Livesey, Sharon M. 2002. Global Warming Wars: Rhetorical and Discourse Analytic Approaches to ExxonMobil's Corporate Public Discourse. *Journal of Business Communication* 39/1, 117-148.

Ljung, Magnus 2002. What Vocabulary Tells Us about Genre Differences: A Study of Lexis in Five Newspaper Genres. In Breivik, Leiv Egil / Hasselgren, Angela (eds) *From the COLT's Mouth ... and Others.* Amsterdam: Rodopi, 181-196.

Locher, Miriam A. / Hoffmann, Sebastian 2006. The Emergence of the Identity of a Fictional Expert Advice-giver in an American Internet Advice Column. *Text & Talk.* 26/1, 69-106.

Locke, Joanne 2004. *A History of Plain Language in the United States Government*. Online at <www.plainlanguage.gov>.

López, Ángeles 2008a. La reprogramación celular, estrella de la investigación en 2008. *WebSalud, El Mundo.* 18 December, 2008.

López, Ángeles 2008b. El riesgo de 'investigar' con lo desconocido. *WebSalud, El Mundo.* 25 September 2006.

López, Xóse / Moisés, Limia / Antonio, Isasi / Pereira, Xosé / Gago, Manuel / Calvo, Rosario 2005. Tipología de los cibermedios. In Salaverria, Ramón. *Cibermedios. El impacto de Internet en los medios de comunicación en España*. Sevilla: Comunicación Social Ediciones y Publicaciones, 39-82.

Lorenzoni, Paulo José / Almada de Souza, Raquel Canzi / Kohara, Suely Keiko / Beenke França, João César / Rodrigues, Giovanna Assis / Rocha de Carvalho, José Gastão 2007. O pôster em encontros científicos. *Revista Brasileira de Educação Médica*. 31/3, 304-309.

Lorés-Sanz, Rosa 2004. On RA Abstracts: From Rhetorical Structure to Thematic Organisation. *English for Specific Purposes*. 23/3, 280-302.

Lorés-Sanz, Rosa 2006. 'I will argue that': First Person Pronouns as Metadiscoursal Devices in Research Article Abstracts in English and Spanish. *ESP Across Cultures*. 3, 23-40.

Lorés-Sanz, Rosa 2008, Genres in Contrast: The Exploration of Writers' Visibility in Research Articles and Research Article Abstracts. In Burgess, Sally / Martín-Martín, Pedro (eds) *English as an Additional Language in Research Publication and Communication.* Bern: Peter Lang, 105-122.

Machin, David / van Leeuwen, Theo 2007. *Global Media Discourse*. London: Routledge.

Maci, Stefania 2010. The Modus Operandi of Litigation in Arbitration. In Bhatia/Candlin/Gotti (eds) 2010, 69-83.

Maci, Stefania 2011. Genre Variation in Medical Discourse: The Case of Medical Posters. In Sarangi/Polese/Caliendo (eds),169-189.

MacIntosh-Murray, Anu 2007. Poster Presentations as a Genre in Knowledge Communication: A Case Study of Forms, Norms, and Values. *Science Communication*. 28/3, 347-376.

Maingueneau, Dominique 2002. Les rapports des organisms internationals: un discours constituant? In Rist, Gilbert (ed.) *Les mots du pouvoir. Sens et non-sens de la rhétorique international*. Paris: PUF, 119-132.

Malavasi, Donatella 2011. 'Doing well by doing good': a comparative analysis of Nokia's and Ericsson's Corporate Social Responsibility Reports. In Garzone, Giuliana / Gotti, Maurizio (eds) *Discourse, Communication and the Enterprise. Genres and Trends*. Bern: Peter Lang, 193-212.

Marriott, Arthur 2004. Less is More: Directing Arbitration Procedures. *Arbitration International*. 16/3, 353-56.

Martin, James R. 1993. Genre and Literacy – Modelling Context in Educational Linguistics. *Annual Review of Applied Linguistics*. 13, 141-172.

Martin, James R. 2002. A Universe of Meaning. How Many Practices? In Johns, Ann M. (ed.) *Genre in the Classroom. Multiple Perspective*. Mahwah (N.J.): Lawrence Erlbaum Associates, 269-278.

Martinet, André 1955. *Économie des changements phonétiques*. Bern: Francke.

Martín-Martín, Pedro 2003. A Genre Analysis of English and Spanish Research Paper Abstracts in Experimental Social Sciences. *English for Specific Purposes*. 22/1, 25-43.

Marwick, Alice, E. / Boyd, Danah 2010. I Tweet Honestly, I Tweet Passionately: Twitter Users, Context Collapse, and the Imagined Audience. *New Media & Society*. 13/1, 114-133.

Matthews, Janice R. / Matthews, Robert W. 1996. *Successful Scientific Writing. A Step-by-step Guide for the Biological and Medical Sciences*. Cambridge: Cambridge University Press.

Mattli, Walter 2001. Private Justice in a Global Economy: From Litigation to Arbitration. *International Organization*. 554/4, 919-947.

Maugh, Thomas H. 1974. Speaking of Science: Poster Sessions: A New Look at Scientific Meetings. *Science*. 184/4144, 1361.

McCarthy, Michael 2001. *Issues in Applied Linguistics*. Cambridge: Cambridge University Press.

McIntosh, Andrew 1991. The Impact of Environmental Issues on Marketing and Politics in the 1990s. *Journal of the Market Research Society*. 33/3, 205-217.

Melander, Björn / Swales, John M. / Fredrickson, Kirstin M. 1997. Journal Abstracts from Three Academic Fields in the United States and Sweden: National or Disciplinary Proclivities? In Duszak, Anna (ed.) *Culture and Styles of Academic Discourse*. Berlin: Mouton de Gruyter, 251-272.

Miles, Wendy 2003. Practical Issues for Appointment of Arbitrators: Lawyer vs. Non-Lawyer and Sole Arbitrator vs. Panel of Three or More. *Journal of International Arbitration*. 20/3: 219-32.

Miller, Carolyn R. 1984/1994. Genre as Social Action. In Freedman, Aviva / Medway, Peter 1994 (eds). *Genre and the New Rhetoric*. London: Francis & Taylor, 23-42 (originally in *Quarterly Journal of Speech*, 1984, 70, 151-167).

Miller, Jane E. 2007. Preparing and Presenting Effective Research Posters. *HSR: Health Services Research*. 42/1, 311-328.

Mitra, Ananda / Cohen, Elisia 1999. Analyzing the Web: Directions and Challenges. In Jones, Steve (ed.) *Doing Internet Research: Critical Issues and Methods for Examining the Net*, Thousand Oaks (CA): Sage, 179-202.

Mitrany, Devora 2005. Creating Effective Poster Presentations: The Editor's Role. *Science Editor*. 28/4, 114–116.

Morales, Pol 2007. Vladimir de Semir: 25 años de periodismo científico. En *Cuadernos de la Fundación Dr. Esteve*. 11, XIII-XIX.

Morris, Joan M. 1988. Images of Community in Scientific Texts – A Grid Group-Approach to the Analysis of Journal Editorials. *Sociological Inquiry*. 58/3, 240-260.

Motta-Roth, Désirée 1996. Same Genre, Different Discipline: A Genre-Based Study of Book Reviews in Academe. *The ESPecialist*. 27/2, 99-131.

Müller, Simone 2005. *Discourse Markers in Native and Non-native English Discourse*. Amsterdam: John Benjamins.

Myers, Greg 1991. Stories and Styles in Two Molecular Biology Review Articles. In Bazerman, Charles / Paradis, James (eds) *Textual Dynamics of the Professions. Historical and Contemporary Studies of Writing in Professional Communities*. Madison: The University of Wisconsin Press, 45-75.

Nabatchi, Tina 2007. The Institutionalization of Alternative Dispute Resolution in the Federal Government. *Public Administration Review*. 67/4, 646-661.

Nariman, Fali S. 2000. The Spirit of Arbitration: The Tenth Annual Goff Lecture. *Arbitration International*. 16/3: 261-278.

Nelson, Theodor H. 1981. *Literary Machines*. Swarthmore, PA.: self-published.

Newmark, Peter 1988. *A Textbook of Translation*. London: Prentice Hall.

Nickerson, Catherine / de Groot, Elizabeth 2005. Dear Shareholder, Dear Stockholder, Dear Stakeholder: The Business Letter Genre in the Annual General Report. In Gillaerts/Gotti (eds), 325-346.

Norlyk, Birgitte 2006. Clashing Norms: Job Ads or Job Narratives. In Gillaerts, Paul / Shaw, Philip (eds) *The Map and the Landscape. Norms and Practices in Genre.* Bern: Lang, 43-62.

Nye, Joseph S. Jr. 2008. Public Diplomacy and Soft Power. *The Annals of the American Academy of Political and Social Science.* 616/1, 94-109.

O'Reilly, Tom 2005. What is Web 2.0. Design Patterns and Business Models for the Next Generation of Software. Online at <http://oreilly.com/web2/archive/what-is-web-20.html>.

Orlikowski, Wanda J. / Yates, JoAnne 1994. Genre Repertoire: Norms and Forms for Work and Interaction. *CCS MIT Working Papers* 166. Available at <http://ccs.mit.edu/papers/CCSWP166.html>.

Ortner, Lorelies 1992. Textkonstitutive Merkmale von Stellenangeboten um 1900. *Deutsche Sprache.* 20, 1-31.

Ortner, Lorelies 1993. Die Bildung von Berufsbezeichnungen (um 1900 und heute). In Wellmann, Hans (ed.) *Synchrone und diachrone Aspekte der Wortbildung.* Heidelberg: Carl Winter, 29-65.

Ortner, Lorelies 1997. Der Stil der Anzeigen in der Wiener Presse um 1900. In Scheichl, Sigurd P. / Duchkowitz, Wolfgang (eds) *Zeitungen im Wiener Fin de siècle*. Wien/München: Verlag für Geschichte und Politik/Verlag Oldenbourg, 90-122.

Ortner, Lorelies 1999. Stellenanzeigen und Geschlecht. Sprachwandel in österreichischen Zeitungen des 20. Jahrhunderts. In Pümpel-Mader, Maria / Schönher, Beatrix (eds) *Sprache – Kultur – Geschichte. Sprachhistorische Studien zum Deutschen.* Festschrift für Hans Moser zum 60. Geburtstag. Innsbruck: Universität Innsbruck (Innsbrucker Beiträge zur Kulturwissenschaft. Germanistische reihe 59), 325-359.

Palacios, Marcos / Díaz, Javier 2009. *Online Journalism: Research Methods. A Multidisplinary Approach in Comparative Perspective*. Bilbao: Servicio Editorial de la Universidad del País Vasco.

Palmer, Frank R. 2001. *Mood and Modality. Second Edition.* Cambridge: Cambridge University Press ([1]1986).

Paltridge, Brian 1996. Genre, Text Type, and the Language Learning Classroom. *ELT Journal.* 50, 237-243.

Paltridge, Brian 1999. Reading across the Curriculum: A Genre-based Perspective. In Short, Deborah (ed.) *New Ways in Teaching English at the Secondary Level.* Alexandra VA: TESOL.

Paltridge, Brian 2001. *Genre and the Language Learning Classroom*. Ann Arbor, USA: University of Michigan Press.

Paré, Antohny 2002. Genre and Identity: Individuals, Institutions, and Ideology. In Coe, Richard / Lingard, Lorelei / Teslenko, Tatiana (eds). *The Rhetoric and Ideology of Genre: Strategies for Stability and Change*. Cresskill, New Jersey: Hampton Press, 57-71.

Paul, Hermann 1920. *Prinzipien der Sprachgeschichte*. Halle: Niemeyer ([1]1880).

Pennarola, Cristina 2003. *Nonsense in Advertising. 'Deviascion' in English Print Ads*. Naples: Liguori.

Perelman, Chaïm 1982. *The Realm of Rhetoric*, transl. W. Kluback. Notre Dame: University of Notre Dame Press.

Perelman, Chaïm / Olbrechts-Tyteca, Lucie 1969. *The New Rhetoric: A Treatise on Argumentation*, transl. J. Wilkinson / P. Weaver. Notre Dame: University of Notre Dame Press.

Pho, Phuong D. 2008. Research Article Abstracts in Applied Linguistics and Educational Technology: A Study of Linguistic Realizations of Rhetorical Structure and Authorial Stance. *Discourse Studies*. 10/2, 231-250.

Polese, Vanda 2006. Patterning, Regularity and Variation in the EUSLEX Corpus on Line. *Linguistica e Filologia*. 23, 91-111.

Polese, Vanda / D'Avanzo, Stefania 2010. Linguistic and Legal Vagueness in EU Directives Harmonising Protection for Refugees and Displaced Persons. In Giannoni, Davide S. / Frade, Celina (eds) *Researching Language and the Law: Textual Features and Translation Issues*. Bern: Peter Lang, 89-111.

Polese, Vanda / D'Avanzo, Stefania 2011. Linguistic and Legal Vagueness in EU Directives on Immigration and Asylum: The Value(s) of *Should*. In Di Martino, Gabriella / Lombardo, Linda / Nuccorini, Stefania (eds) *Challenges for the 21st Century: Dilemmas, Ambiguities, Directions*. Vol. 2: Language Studies. Roma: Edizioni Q, 397-405.

Polese, Vanda / D'Avanzo, Stefania (forth.) Popularization and Legal Genre(s): EU Summaries of Directives on Migrants' Rights. In Caliendo, Giuditta / Bongo, Giancarmine (eds) *The Language of Popularization: Theoretical and Descriptive Models*. Bern: Peter Lang.

Politis, Periclis / Kakavoulia, Maria 2006. Direct Discourse in the Greek Press: From Evidentiality to Subjectivity. *Revista Alicantina de Estudios Ingleses*. 19, 345-363.

Pubantz, Jerry 2005. Constructing Reason: Human Rights and the Democratization of the United Nations. *Social Forces*. 84/2, 1291-1302.

Quintarelli, Emanuele 2005. Folksonomies: Power to the People, Paper presented at the ISKO Italy-UniMIB meeting. Milan, 24 June 2005. <http://www.iskoi.org/doc/folksonomies.htm>.

Ragonese, Ruggero 2010. Guide turistiche. Un'introduzione. *Rivista on-line dell'Associazione Italiana di Studi Semiotici.* IV/6, 5-18.

Ramus, Catherine A. / Montiel, Ivan 2005. When are Corporate Environmental Policies a Form of Greenwashing? *Business and Society*. 44/4, 377-414.

Rensberger, Boyce 2009. Science Journalism: Too Close for Comfort. *Nature*, 459, 1055-1056. Available at <http://www.nature.com/nature/journal/v459/n7250/full/4591055a.html>.

Riley, Philip 2006. Self-expression and the Negotiation of Identity in a Foreign Language. *International Journal of Applied Linguistics.* 16/3, 295-318.

Rimmon-Kenan, Shlomith 1983. *Narrative Fiction. Contemporary Poetics.* London: Routledge.

Rist, Gilbert 2002. Le prix des mots. In Rist, Gilbert (ed.) *Les mots du pouvoir. Sens et non-sens de la rhétorique international.* Paris: PUF, 9-23.

Salager-Meyer, Françoise 2002. Market-Place, Self-Confidence and Criticism in Medical Editorials. *Revista Canaria de Estudios Ingleses*. 44, 65-78.

Santulli, Francesca 2007. Il discorso procedurale come tratto distintivo della guida turistica: evoluzione diacronica e variazioni sincroniche. In Garzone, Giuliana / Salvi, Rita (eds). *Linguistica, linguaggi specialistici, didattica delle lingue. Studi in onore di Leo Schena.* Roma: CISU, 227-240.

Sarangi, Srikant / Polese, Vanda / Caliendo, Giuditta (eds) 2011. *Genre(s) on the Move. Hybridization and Discourse Change in Specialized Communication*. Naples: Edizioni Scientifiche Italiane (ESI).

Scollon, Ron 2008. *Analyzing Public Discourse*. London: Routledge.

Scollon, Ron / Scollon, Suzanne Wong 1995. *Intercultural Communication: A Discourse Approach*. Oxford: Blackwell.

Scott, Marvin / Lyman, Stanford 1968. Accounts. *American Sociological Review*. 33, 46-62.

Scott, Mike 1999a, *WordSmith Tools version 3.0*, Oxford: Oxford University Press.

Scott, Mike 1999b, *WordSmith Tools version 3.1*, Oxford: Oxford University Press.

Scott, Mike 1999c. *WordSmith Tools Help Manual*. Version 3.0. Oxford: Mike Scott & Oxford University Press.

Scott, Mike 2004. *Wordsmith Tools 4.0*, Oxford: Oxford University Press.

Scott, Mike 2007. *WordSmith Tools 5.0*. Oxford: Oxford University Press.

Senarclens, Pierre de 2002. Contraintes politiques et institutionnelles du discours des Nations Unies. In Rist, Gilbert (ed.) *Les mots du pouvoir. Sens et non-sens de la rhétorique internationale.* Paris: PUF, 133-143.

Simpson, Paul 1993. *Language, Ideology and Point of View.* London: Routledge.

Sinclair, John M. 1986. Fictional Worlds. In Coulthard, Malcom (ed.) *Talking about Text*. Birmingham: University of Birmingham ELR, 43-60.

Sinclair, John M. 1988. Mirror for a Text. *The Journal of English and Foreign Language*. University of Hyderabad: India. 1, 15-45.

Skjelsbaik, Kjell 1991. The UN Secretary-General and the Mediation of International Disputes. *Journal of Peace Research.* 28/1, 99-115.

Smith, Philip E. M. / Fuller, Geraint / Dunstan, Frank 2004. Scoring Posters at Scientific Meetings: First Impressions Count. *Journal of the Royal Society of Medicine*. 97, 340-341.

Solly, Martin 2002. 'Once a Trademark, not Always a Trademark': Using Language to Avoid Legal Controversy. In Gotti, Maurizio / Heller, Dorothee / Dossena, Marina (eds) *Conflict and Negotiation in Specialized Texts*. Bern: Peter Lang, 211-232.

Solly, Martin 2011. Setting the Graduates Challenges for the 21st Century: The Case of Commencement Speeches. In Di Martino, Gabriella / Lombardo, Linda / Nuccorini, Stefania (eds) *Challenges for the 21st Century. Dilemmas, Ambiguities, Directions. Papers from XXIV AIA Conference*. *Vol. II Language Studies.* Rome: Edizioni Q, 357-365.

Sosnoski, James 1999. Configuring as a Mode of Rhetorical Analysis. In Jones, Steven (ed.) *Critical Issues and Methods for Examining the Net*. Thousand Oaks (CA): Sage, 127-143.

Spence, Crawford / Gray, Robert H. 2007. *Social and Environmental Reporting and the Business Case*. ACCA Research Report 98. London: ACCA.

Stanojeviæ-Bakiæ, Nevenka 2002. Poster Presentation. *Archive of Oncology*. 10/3, 213-4.

Stöckl, Hartmut 2004. In between Modes: Language and Image in the Printed Media. In Ventola, Eija / Charles, Cassily / Kaltenbacher, Martin (eds), *Perspectives on Multimodality*. Amsterdam/Philadelphia: John Benjamins, 9–30.

Stotesbury, Hikka 2003. Evaluation in Research Article Abstracts in the Narrative and Hard Sciences. *Journal of English for Academic Purposes*. 2/4, 327-341.

Swales, John M. 1990. *Genre Analysis: English in Academic and Research Settings.* Cambridge: Cambridge University Press.

Swales, John M. 1996. Occluded Genres in the Academy: The Case of the Submission Letter. In Ventola, Eija / Mauranen, Anna (eds) *Academic Writing. Intercultural and Textual Issues*. Amsterdam: John Benjamins, 45-58.

Swales, John M. 2004. *Research Genres: Explorations and Applications*. New York: Cambridge University Press.

Swales, John M. / Feak, Christine 2000. *English in Today's Research World: A Writing Guide*. Ann Arbor: University of Michigan Press.

Timmerman, P.E. 1992. *Handboek voor de Personeelsadvertentie*. Groningen: Boekwerk.

Toolan, Michael 1988. *Narrative. A Critical Linguistic Introduction*. London: Routledge.

United Nations 2000. *We the Peoples: The Role of the United Nations in the Twenty-First Century. Report of the Secretary-General*. New York: United Nations. <http://www.un.org/millennium/sg/report/>.

United Nations 2007. *Writing for the United Nations: An Online Learning Programme*. <http://69.94.137.26/editorialcontrol/training/Index.htm>.

Vagle, Wenche 2001. The Diachronicity of Text Norms. A Theory of Genre Formation and Change. In Berge, Kjell Lars (ed.) *Teksthistorie – Tekstvitenskapelige Bidrag*. Oslo: Norsk Sakprosa, 141-214.

Valerio, María 2005. Nueva alternativa para obtener células madre sin destruir embriones. *WebSalud, El Mundo*. 17 October, 2005.

Valerio, María 2009. Una nueva técnica celular frena la herencia de enfermedades genéticas en monos. *WebSalud, El Mundo*. 26 August, 2009.

van Bonn, Sarah / Swales, John 2007, English and French Journal Abstracts in the Language Sciences: Three Exploratory Studies. *Journal of English for Academic Purposes*. 6/2, 93-108.

van Deuren, M. / Lievens, S. 1987/1989. *De 'ideale' personeelsadvertentie. Inhoudelijke analyse van 2104 personeelsadvertenties met betrekking tot personeelsbeleid en studieadvies*. Dossier Personeelsbeleid, nr 2. 1989. Brussel: Ced. Samsom.

van Diik, Teun A. 2005. Contextual knowledge management in discourse production. A CDA perspective. In Wodak, Ruth / Chilton, Paul (eds) *A New Agenda in (Critical) Discourse Analysis*. Amsterdam: Benjamins, 71-100.

van Eemeren, Frans H. / Grootendorst, Rob 1992. *Argumentation, Communication, and Fallacies: A Pragma-dialectical Perspective*. Hillsdale, NJ: Lawrence Erlbaum Associates.

van Eemeren, Frans H. / Houtlosser, Peter 1999. Strategic Manoeuvring in Argumentative Discourse. *Discourse Studies*. 1/4, 479-497.

van Eemeren, Frans H. / Houtlosser, Peter 2002. Strategic Manoeuvring with the Burden of Proof. In Eemeren, Frans H. van (ed.) *Advances in Pragma-dialectics*. Amsterdam: SicSat / Newport News, VA: Vale Press, 13-28.

van Eemeren, Frans H. / Houtlosser, Peter 2006. Strategic Maneuvering: A Synthetic Recapitulation. *Argumentation*. 20/4, 381-392.

van Eemeren, Frans H / Houtlosser, Peter / Snoeck-Henkemans, Francisca A. 2007. *Argumentative Indicators in Discourse*. Amsterdam: Springer.

van Meurs, Frank / Korzilius, Hubert / Den Hollander, Adriënne 2005. Testing the Effect of a Genre's Form on its Target Group. In Gillaerts, Paul / Shaw, Philip (eds) *The Map and the Landscape. Norms and Practices in Genre*. Bern: Peter Lang, 91-114.

Verschueren, Jef 1999. *Understanding Pragmatics*. London: Arnold.

Virtanen, Tuija / Halmari, Helena. 2005. Persuasion across Genres. Emerging Perspectives. In Halmari /Virtanen (eds), 3-24.

Wang, Daohuan 2009. 科普作為一種志業：使命、責任、實踐 (Popular Science as a Vocation: Tasks, Responsibilities, & Practice). Speech presented at National Normal Taiwan University, Taipei (video). <http://case.ntu.edu.tw/SciMonthly40/sm40blog/?p=583>.

Ware, B.L. / Linkugel, Wil A. 1973. They Spoke in Defense of Themselves: On the Generic Criticism of Apologia. *Quarterly Journal of Speech*. 59, 273-283.

Weinreich, Uriel 1954. Is a Structural Dialectology Possible? *Word. Journal of the Linguistic Circle of New York*. 10, 388-400.

Weinreich, Uriel / Labov, William / Herzog, Marvin 1968. Empirical Foundations for a Theory of Language Change. In Lehmann, Winfred / Malkiel, Yakov (eds) *Directions for Historical Linguistics. A Symposium*. Austin: University of Texas Press, 95-195.

Werlich, Egon 1975. *Typologie der Texte*. Heidelberg: Quelle und Meyer.

Werlich, Egon 1983. *A Text Grammar of English*. Heidelberg: Quelle and Meyer ([1]1976).

Whimster, William F. 1989. Wanted: Reader Friendly Posters. As Much Show Business as Science. *British Medical Journal*. 298, 274.

White, Allen L. 1999. Sustainability and the Accountable Corporation: Society's Rising Expectations of Business. *Environment*. 41/8, 30-43.

Whitley, Richard 1985. Knowledge Producers and Knowledge Acquirers: Popularisation as a Relation between Scientific Fields and Their Publics. In Shinn, Terry / Whitley, Richard (eds) *Expository Science: Forms and Functions of Popularisation. Sociology of the Sciences Yearbook* Vol. 9. Dordrecht and Boston: Reidel, 3-28.

Willett, Lisa L. / Paranjape, Anuradha / Estrada, Carlos 2008. Identifying Key Components for an Effective Case Report Poster: An Observational Study. *Journal of General Internal Medicine.* 24/3, 393-397.

Williams, Christopher 2005/2007. *Tradition and Change in Legal English.* Bern: Peter Lang.

Williams, Christopher 2006. Fuzziness in Legal English: What Shall We Do with 'Shall'? In Wagner, Anne / Cacciaguidi-Fahy, Sophie (eds) *Legal Language and the Search for Clarity*. Bern: Peter Lang, 237-263.

Wodak, Ruth 2009. *The Discourse of Politics in Action.* London: Palgrave Macmillan.

Yates, JoAnne / Orlikowski, Wanda 1992. Genres of Organizational Communication: A Structural Approach. *Academy of Management Review.* 17: 299-326.

Yates, JoAnne 1989. *Control through Communication: The Rise of System in American Management.* Baltimore: Johns Hopkins University Press.

Yates, JoAnne 2005. *Structuring the Information Age: Life Insurance and Information Technology in the 20th Century.* Baltimore: Johns Hopkins University Press.

Young, David 1988. Projection and Deixis in Narrative Discourse. In Coupland, Nicholas (ed.) *Styles of Discourse*. London: Croom Helm, 20-49.

Zobell, Karlo 2002. *Rekruteringscommunicatie: Strategie, Middelen, Technnieken.* Mechelen: Kluwer.

Notes on Contributors

MARINA BONDI is Professor of English Language and Translation at the University of Modena and Reggio Emilia, Italy, where she is currently Dean of the Faculty of Letters and Philosophy. She has published on various aspects of discourse analysis and EAP, with particular reference to the argumentative features of academic discourse and the role of metadiscourse and evaluative language. Her recent work centres on language variation across genres, disciplines and cultures through the analysis of small specialized corpora, with an emphasis on the humanities and social sciences. On language variation in specialized discourse she has published a number of articles and co-edited a number of volumes: most recently, *Academic Discourse Across Disciplines* with Ken Hyland, *Managing Interaction in Professional Discourse: Intercultural and Interdiscoursal Perspectives* with Julia Bamford and *Keyness in Texts* with Mike Scott.

PAOLA CATENACCIO, is Associate Professor of English Linguistics and Translation at Università degli Studi di Milano. Her research interests span from second language acquisition to translation and discourse analysis, particularly in the context of LSP research. Her most recent contributions focus on linguistic aspects of corporate communication, especially corporate social responsibility. She has authored numerous articles which have appeared in international journals and edited collections. Some of her latest publications are: “Social and environmental reports. A diachronic perspective on an emerging genre” (2011); “Green advertising and corporate CSR communication: hybrid discourses, mutant genres” (2011); *Corporate Press Releases: An Overview* (2008); “Press releases as a hybrid genre: Addressing the informative/promotional conundrum” (2008); “De-humanising the alien: The construction of migrants’ rights in EU legislation” (2007). She has also co-edited, with Giuliana Garzone and Gina Poncini, *Multimodality in Corporate Communication. Web Genres and Discursive*

Identity (2007) and, with Giuliana Garzone, *Language and Bias in Specialised Discourse* (2008) and *Identities Across Media and Modes: Discursive Perspectives* (2009).

GERMANA D'ACQUISTO is a secondary school teacher and a lecturer of English language and linguistics at the Faculty of Law and Faculty of Arts, University of Naples 'Federico II'. She was awarded a Doctorate in English for Special Purposes by the University of Naples 'Federico II' discussing a thesis on "The language of the UN Resolution on the Question of Palestine". She has researched on institutional discourse within the national research Project PRIN 2007. Her recent work deals with intercultural language teaching and aspects of modernization of legal language. She is a member of ENIEDA (European Network for Intercultural Education Activities). She has published the following research articles: "Italy and xenophobia: the case of gypsies and Italian attitude towards foreign people" in *Jolie Journal of Linguistic and Intercultural Education* (2008); "Exploring Language Behaviour in ESP courses" with S. D'Avanzo and V. Polese in *Plurilinguismo e integrazione: abilità e competenze linguistiche in una società multietnica* (2010).

CHIARA DEGANO is a Tenured Researcher in English Linguistics and Translation at Università degli Studi di Milano, Italy. Her research is centred on Discourse Analysis, which she has integrated with the quantitative approach of corpus linguistics and with aspects of Argumentation theory. Within this framework she has pursued a twofold line of research, focusing on the representation of international conflicts in the media (*Discorsi di guerra: il prologo del conflitto iracheno sulla stampa britannica e italiana*, 2008; "Rhetoric and bias in Editorials: the Lebanese crisis of 2006 in the British press", 2008, with G. Garzone; "Presupposition and dissociation in discourse: a corpus study", 2007), and on professional discourse, which she investigated in the perspective of ideological and evaluative elements, with specific reference to issues of identity ("The construction of cultural identity in banks' reports on corporate social responsibility", 2010). She has also taken part in an international project on arbitration, investigating generic tensions in international awards. Her other research interests fall in the area of translation and interference, with a concern for syntactic interference in translations from English into Italian and for the rendering of multi-cultural references in film dubbing.

Sylvain Dieltjens (Ph.D.) teaches French and business communication in the Department of Applied Language Studies of Lessius Antwerp and is affiliated researcher at the Catholic University of Louvain. His research focuses on linguistic and strategic aspects of business communication and of palliative discourse. Together with Priscilla Heynderickx he has published several articles in journals such as *Ad Rem, Document Design, The Journal of Technical Writing & Communication, Onze Taal* and *Tekst[blad]*. He has co-edited a volume on *Displaying Competence* (with Chris Braecke and Geert Jacobs). In addition, he is president of the communication agency TxT-Ibis nv, Kortenberg (<www.txt-ibis.be>) and President of the Vlaamse Vereniging voor Zakelijke Communicatie, a Belgian organization of business communication professionals and researchers (<www.vvzc.be>).

Stefania D'Avanzo is a Lecturer in the Faculty of Foreign Languages and Literatures at the University of Naples 'L'Orientale' and in the Faculty of Economics at the University of Sannio (Benevento). She is also a teaching assistant in the Faculty of Law at the University of Naples 'Federico II'. She holds a Doctoral degree in English for Special Purposes and a Master's degree in Linguistics and Sociolinguistics of European Languages from the University of Naples 'Federico II'. She has presented her ongoing research at international seminars and conferences. Her current research mainly focuses on legal and institutional discourse, academic discourse, discourse analysis and second language acquisition.

Paula de Santiago González is an Assistant Professor in the Department of English Studies at the University of Valladolid. She graduated from the University of Valladolid in English Studies. She has got two MAs in TESOL/Linguistics from West Virginia University and in Translation and Interpreting from the University of Valladolid. In 2012 she defended her PhD dissertation, entitled *Estudio intra e interlingüístico de la variación terminológica en el lenguaje de la biomedicina: las células madre*. She has been a visiting professor/researcher at West Virginia University, Università degli Studi di Milaon and at the University of Ottawa. De Santiago's main research is on terminology and specialized phraseology in different registers and textual genres. In this field, she has organized and participated in several national and international conferences in Europe and North America. She has coauthored the *Diccionario terminológico y fraseólogico español/inglés sobre vitivinicultura* (2009), and the study "El mapa conceptual en el glosario Linguaturismo. Metodología de elaboración y aplicaciones" (with Elena Landone), in *La lengua del turismo. Géneros discursivos y terminología* (2011). She is a member of the ACTRES research group which works on corpus-based studies; it is coordinated by Prof. Rosa Rabadán at the University of Leon.

GIULIANA GARZONE is Full Professor of English Linguistics and Translation at Università degli Studi di Milano, Italy. Her main field of investigation is specialized discourse in its different domains, which she has explored in a discourse analytical perspective, integrating it whenever necessary with computerized interrogation routines, also in light of her work on the theoretical issues involved in combining Critical Discourse Analysis with Corpus Linguistics ("What can corpus linguistics do for Critical Discourse Analysis?", 2004, with F. Santulli). She has co-ordinated several research projects and published extensively on legal language (*Performatività e linguaggio giuridico: una proposta di classificazione*, 1999; *Il linguaggio giuridico. Prospettive interdisciplinari*, 2008, with F. Santulli), on scientific discourse (*Perspectives on ESP and Popularization*, 2006), and on business communication. In the latter domain, she has recently devoted specific attention to multimodality and to the analysis of text types and genres embedded in the web-mediated environment ("Genres, multimodality and the World-Wide Web: Theoretical Issues", 2007), an issue she has also investigated with reference to other types of communication, e.g. political communication (cf. "Public Diplomacy, Multimodality and The World-Wide Web", with C. Degano). She has also (co-)edited several volumes (most recently, *Discourse, Ideology and Specialized Communication*, with S. Sarangi, 2007; *The Use of English in Institutional and Business Settings,* with C. Ilie, 2007; *Dis-*

course and Identities in Institutional, Professional and Academic Communication, with J. Archibald, 2010; *Discourse, Communication and the Enterprise: Genres and Trends*, 2011, with M. Gotti). Her other research interests are in translation and interpreting studies, on which she has also published extensively (among others: *Interpreting in the 21st Century,* with M. Viezzi, 2002; *Esperienze del tradurre*, 2005). She is general editor (with A. Cardinaletti e F. Frasnedi) of the book series *Lingua, traduzione, didattica.*

DAVIDE S. GIANNONI is Associate Professor of English (Language and Linguistics) at the University of Bergamo, Italy, whose Centre for Research on Languages for Specific Purposes (CERLIS) he helped to establish in 1999. He holds a PhD in Applied Linguistics from the University of Reading, an M.A. in Linguistics (TESOL) from the University of Surrey and is a former Morley Scholar of the University of Michigan's English Language Institute. His research interests focus on the linguistic analysis of English academic, legal and business discourse, approached both synchronically and contrastively. He is co-editor of *Language Studies Working Papers* and assistant editor of *The European English Messenger.* In recent years he has acted as a reviewer for a number of peer-reviewed journals (*English for Specific Purposes; Journal of English for Academic Purposes; Ibérica; Issues in Applied Linguistics; Journal of the American Society for Information Science and Technology)* as well as assisting in the organisation of several international conferences in Italy and abroad.

CINZIA GIGLIONI obtained her PhD from the Università degli Studi di Milano in 2002 and received a post-doc research grant from the same university from 2004 to 2008. She has been teaching English Language over the last ten years and at the moment she teaches at Milan University and Bergamo University. Her research interests include genre studies and corporate communication, on which she has presented several papers in national and international conferences. She has published articles focusing on the rhetorical strategies deployed in annual company reports. Since 2005 she has been collaborating with other scholars on the Italian section of ABELL (Annual Bibliography of English Language and Literature) edited by the Modern Humanities Research Association, Cambridge.

PAUL GILLAERTS (Ph.D.) is professor of Dutch at the Department for Applied Linguistics of Lessius, Antwerp (Belgium) and affiliated researcher at the Catholic University of Leuven (Belgium). He teaches Dutch applied linguistics courses on genre, style and text revision. His research concerns applied genre analysis of Dutch texts in professional, academic, institutional and business settings. His research has been published in journals such as *Tijdschrift voor Taalbeheersing* and *Journal of English for Academic Purposes*; he co-edited several volumes on business communciation and genre in the Linguistic Insights series.

PRISCILLA HEYNDERICKX teaches Dutch and business communication at the Department of Applied Linguistics of Lessius University College (Antwerp). She is an affiliated researcher of Catholic University of Leuven and chief editor of *Ad Rem*, a business communication journal (www.vvvz.be). With Sylvain Dieltjens she has published several articles on aspects of business communication in *Ad Rem, Document Design, The Journal of Technical Writing & Communication, Onze Taal, Tekst[blad]*, etc. She has also presented papers on similar subjects at conferences organised by Anéla, the Association for Business Communication, DICOEN, Document Design, VIOT, etc.

STEFANIA M. MACI is a Tenured Researcher of English Language and Translation at the University of Bergamo. She is currently involved in national and international research projects on academic, legal and tourism discourses. Her major publications focus on aspects related to metadiscoursive and argumentative strategies of English academic texts across genres and cultures. Amongst her recent publications are *The Language of Tourism* (2010), and the following articles: “Genre variation in medical discourse: the case of medical posters” (2011), “The *Modus Operandi* of Litigation in Arbitration” (2010), “The Argumentative Structure in American and Italian Medical Research Articles” (2010), “Litigation Procedures in Arbitral Practice: A case of Arbitral Litigation?” (2009); “The Migration of Scientific Knowledge into Alternative Forms of Research Articles" (2009), and *Identity Traits in English Academic Discourse* (2008) co-edited with Davide Giannoni.

MIN-HSIU LIAO is a lecturer in the Department of Languages and Intercultural Studies at Heriot-Watt University in Edinburgh, UK, where she teaches Chinese and English-Chinese translation and interpreting. Her research interests focus on interaction in written communication, and shifts of interaction in the process of translation. She has published on the topic of interaction in the genres of popular science and museum exhibitions.

CRISTINA PENNAROLA is Associate Professor of English Language and Linguistics at the Faculty of Political Science, University of Naples 'Federico II'. She has published a book analysing advertising discourse in the light of nonsense strategies, *Nonsense in Advertising* (2009), and a number of short essays on her research topics ranging from language learning to humanitarian discourse: "An Overview of English for Military Purposes: Linguistic and Professional Issues" (2010); "Of B's and Stings: The Bush, Blair and Berlusconi Jokes across English and Italian", in *TEXTUS, Special Issue on Humour* (2008); "How EU Secondary Legislation Encodes Humanitarian Aid Policies". Her current research interests include institutional discourse. She is a member of the Doctoral Committee in English for Special Purposes at the University of Naples 'Federico II', and she is on the editorial board of the bilingual journal *La Freccia e il Cerchio / The Arrow and the Circle.*

VANDA POLESE is Associate Professor of English Linguistics and Translation in the Faculty of Poltical Science and Lecturer of English for Law in the Faculty of Law at the University of Naples 'Federico II' (Italy). She is a member of the Doctoral Board in English for Special Purposes and of the School of "Scienze Psicologiche, Pedagogiche e Linguistiche" (Psychological, Pedagogical and Linguistic Sciences) (University of Naples 'Federico II'). Her main research interests cover discourse in ESP, especially in domains such as legal, institutional, academic, news discourse, critical discourse analysis, pragmatics, translation, genre hybridisation and popularisation, and applied linguistics. She is the author of *Language in the Spotlight: News Manufacturing and Discourse* (2004), co-editor of *'Languaging' and Interculturality in EU Domains* (2005), *Identity and Culture in English Domain-specific Discourse* (2008), and *Genre(s) on the Move: Hybridization and Discourse Change in Specialized Communication* (2011). Her recent publications include studies on institutional-legal language, applied linguistics, and popularisation and legal genre(s).

FRANCESCA SANTULLI is Associate Professor of Linguistics at IULM University, Milan. Her research has focused on various aspects of language and linguistics, ranging from the history of linguistics to philology, from phonetics to language contact. She published many articles and a book (*L'opera di Hermann Paul tra linguistica e filologia* [*Herman Paul's work between linguistics and philology*], 1995) on the theoretical aspects of language change, with specific reference to the fundamental work of German philologist and linguist H. Paul. Her research also focused on phonetics and phonology, with particular emphasis on the didactic implications (*La voce e la macchina. Fonetica, glottodidattica, multimedialità* [*Voices and machines. Phonetics, language teaching, multimedia*], 1999, with G. Garzone). She has also published on translation and interpreting, as well as on interference and borrowing. She has applied linguistic and rhetorical models of analysis to *corpora* of texts belonging to different genres, combining qualitative and quantitative methodologies. In a book on political discourse (*Le parole del potere, il potere delle parole. Retorica e discorso politico* [*Power of words, words of power. Rhetoric and political discourse*], 2005) she integrates different theoretical approaches and presents a detailed analysis of political texts belonging to different sub-genres. Her most recent book (*Pragmatica della comunicazione turistica* [*The Pragmatics of tourism communication*], 2007, with D. Antelmi and G. Held) applies the principles of pragmatic analysis to texts produced in the realm of tourist communication (travel guidebooks, websites, advertisements).

MARTIN SOLLY is Associate Professor of English Language and Linguistics at the University of Florence. His main current research interests concern language learning in higher education and specialised discourse in academic and professional settings. He is particularly concerned with the relationship between language and context (institutional, disciplinary, intercultural etc), as well as with how language choice impacts on the construction and representation of identity. He has co-edited the volumes *Verbal/Visual Narrative Texts in Higher Education* (2008) and *Identity and Culture in English Domain-specific Discourse* (2008). His recent publications include "Implementing the Bologna Process in Italy" in *ESP in European Higher Education* (2008) and "Using language to shape identity" in *Commonality and Individuality in Academic Discourse* (2009).

ANNARITA TAVANI is a secondary school teacher and a PhD student at Università degli Studi di Milano. She is presently working on her dissertation, entitled *Institutional Language and Discourse in International Organizations*, which focuses on institutional discourse and, in particular, on the genre of the annual report drafted by international organizations. She took part in the International ABC Mentorship Programme for the years 2011/2012 with a paper on the generic structure of the WHO reports, tutored by Professor Marinel Gerritsen, from Nijmegen University, Department of Business Communication Studies, and Professor Priscilla Heynderickx, from Lessius.

Linguistic Insights

Studies in Language and Communication

This series aims to promote specialist language studies in the fields of linguistic theory and applied linguistics, by publishing volumes that focus on specific aspects of language use in one or several languages and provide valuable insights into language and communication research. A cross-disciplinary approach is favoured and most European languages are accepted.

The series includes two types of books:

- **Monographs** – featuring in-depth studies on special aspects of language theory, language analysis or language teaching.
- **Collected papers** – assembling papers from workshops, conferences or symposia.

Each volume of the series is subjected to a double peer-reviewing process.

Vol. 1 Maurizio Gotti & Marina Dossena (eds)
Modality in Specialized Texts. Selected Papers of the 1st CERLIS Conference.
421 pages. 2001. ISBN 3-906767-10-8 · US-ISBN 0-8204-5340-4

Vol. 2 Giuseppina Cortese & Philip Riley (eds)
Domain-specific English. Textual Practices across Communities and Classrooms.
420 pages. 2002. ISBN 3-906768-98-8 · US-ISBN 0-8204-5884-8

Vol. 3 Maurizio Gotti, Dorothee Heller & Marina Dossena (eds)
Conflict and Negotiation in Specialized Texts. Selected Papers of the 2nd CERLIS Conference.
470 pages. 2002. ISBN 3-906769-12-7 · US-ISBN 0-8204-5887-2

Vol. 4 Maurizio Gotti, Marina Dossena, Richard Dury, Roberta Facchinetti & Maria Lima
Variation in Central Modals. A Repertoire of Forms and Types of Usage in Middle English and Early Modern English.
364 pages. 2002. ISBN 3-906769-84-4 · US-ISBN 0-8204-5898-8

Editorial address:

Prof. Maurizio Gotti Università di Bergamo, Facoltà di Lingue e Letterature Straniere,
Via Salvecchio 19, 24129 Bergamo, Italy
Fax: 0039 035 2052789, E-Mail: m.gotti@unibg.it

Vol. 5 Stefania Nuccorini (ed.)
Phrases and Phraseology. Data and Descriptions.
187 pages. 2002. ISBN 3-906770-08-7 · US-ISBN 0-8204-5933-X

Vol. 6 Vijay Bhatia, Christopher N. Candlin & Maurizio Gotti (eds)
Legal Discourse in Multilingual and Multicultural Contexts.
Arbitration Texts in Europe.
385 pages. 2003. ISBN 3-906770-85-0 · US-ISBN 0-8204-6254-3

Vol. 7 Marina Dossena & Charles Jones (eds)
Insights into Late Modern English. 2nd edition.
378 pages. 2003, 2007.
ISBN 978-3-03911-257-9 · US-ISBN 978-0-8204-8927-8

Vol. 8 Maurizio Gotti
Specialized Discourse. Linguistic Features and Changing Conventions.
351 pages. 2003, 2005.
ISBN 3-03910-606-6 · US-ISBN 0-8204-7000-7

Vol. 9 Alan Partington, John Morley & Louann Haarman (eds)
Corpora and Discourse.
420 pages. 2004. ISBN 3-03910-026-2 · US-ISBN 0-8204-6262-4

Vol. 10 Martina Möllering
The Acquisition of German Modal Particles. A Corpus-Based Approach.
290 pages. 2004. ISBN 3-03910-043-2 · US-ISBN 0-8204-6273-X

Vol. 11 David Hart (ed.)
English Modality in Context. Diachronic Perspectives.
261 pages. 2003. ISBN 3-03910-046-7 · US-ISBN 0-8204-6852-5

Vol. 12 Wendy Swanson
Modes of Co-reference as an Indicator of Genre.
430 pages. 2003. ISBN 3-03910-052-1 · US-ISBN 0-8204-6855-X

Vol. 13 Gina Poncini
Discursive Strategies in Multicultural Business Meetings.
2nd edition. 338 pages. 2004, 2007.
ISBN 978-3-03911-296-8 · US-ISBN 978-0-8204-8937-7

Vol. 14 Christopher N. Candlin & Maurizio Gotti (eds)
Intercultural Aspects of Specialized Communication.
2nd edition. 369 pages. 2004, 2007.
ISBN 978-3-03911-258-6 · US-ISBN 978-0-8204-8926-1

Vol. 15 Gabriella Del Lungo Camiciotti & Elena Tognini Bonelli (eds)
Academic Discourse. New Insights into Evaluation.
234 pages. 2004. ISBN 3-03910-353-9 · US-ISBN 0-8204-7016-3

Vol. 16 Marina Dossena & Roger Lass (eds)
Methods and Data in English Historical Dialectology.
405 pages. 2004. ISBN 3-03910-362-8 · US-ISBN 0-8204-7018-X

Vol. 17 Judy Noguchi
The Science Review Article. An Opportune Genre in
the Construction of Science.
274 pages. 2006. ISBN 3-03910-426-8 · US-ISBN 0-8204-7034-1

Vol. 18 Giuseppina Cortese & Anna Duszak (eds)
Identity, Community, Discourse. English in Intercultural Settings.
495 pages. 2005. ISBN 3-03910-632-5 · US-ISBN 0-8204-7163-1

Vol. 19 Anna Trosborg & Poul Erik Flyvholm Jørgensen (eds)
Business Discourse. Texts and Contexts.
250 pages. 2005. ISBN 3-03910-606-6 · US-ISBN 0-8204-7000-7

Vol. 20 Christopher Williams
Tradition and Change in Legal English. Verbal Constructions in Prescriptive Texts.
2nd revised edition. 216 pages. 2005, 2007. ISBN 978-3-03911-444-3.

Vol. 21 Katarzyna Dziubalska-Kolaczyk & Joanna Przedlacka (eds)
English Pronunciation Models: A Changing Scene.
2nd edition. 476 pages. 2005, 2008. ISBN 978-3-03911-682-9.

Vol. 22 Christián Abello-Contesse, Rubén Chacón-Beltrán, M. Dolores López-Jiménez & M. Mar Torreblanca-López (eds)
Age in L2 Acquisition and Teaching.
214 pages. 2006. ISBN 3-03910-668-6 · US-ISBN 0-8204-7174-7

Vol. 23 Vijay K. Bhatia, Maurizio Gotti, Jan Engberg & Dorothee Heller (eds)
Vagueness in Normative Texts.
474 pages. 2005. ISBN 3-03910-653-8 · US-ISBN 0-8204-7169-0

Vol. 24 Paul Gillaerts & Maurizio Gotti (eds)
Genre Variation in Business Letters. 2nd printing.
407 pages. 2008. ISBN 978-3-03911-681-2.

Vol. 25 Ana María Hornero, María José Luzón & Silvia Murillo (eds)
Corpus Linguistics. Applications for the Study of English.
2nd printing. 526 pages. 2006, 2008. ISBN 978-3-03911-726-0

Vol. 26 J. Lachlan Mackenzie & María de los Ángeles Gómez-González (eds)
Studies in Functional Discourse Grammar.
259 pages. 2005. ISBN 3-03910-696-1 · US-ISBN 0-8204-7558-0

Vol. 27 Debbie G. E. Ho
Classroom Talk. Exploring the Sociocultural Structure of Formal ESL Learning.
2nd edition. 254 pages. 2006, 2007. ISBN 978-3-03911-434-4

Vol. 28 Javier Pérez-Guerra, Dolores González-Álvarez, Jorge L. Bueno-Alonso & Esperanza Rama-Martínez (eds)
'Of Varying Language and Opposing Creed': New Insights into Late Modern English.
455 pages. 2007. ISBN 978-3-03910-788-9

Vol. 29 Francesca Bargiela-Chiappini & Maurizio Gotti (eds)
Asian Business Discourse(s).
350 pages. 2005. ISBN 3-03910-804-2 · US-ISBN 0-8204-7574-2

Vol. 30 Nicholas Brownlees (ed.)
News Discourse in Early Modern Britain. Selected Papers of CHINED 2004.
300 pages. 2006. ISBN 3-03910-805-0 · US-ISBN 0-8204-8025-8

Vol. 31 Roberta Facchinetti & Matti Rissanen (eds)
Corpus-based Studies of Diachronic English.
300 pages. 2006. ISBN 3-03910-851-4 · US-ISBN 0-8204-8040-1

Vol. 32 Marina Dossena & Susan M. Fitzmaurice (eds)
Business and Official Correspondence: Historical Investigations.
209 pages. 2006. ISBN 3-03910-880-8 · US-ISBN 0-8204-8352-4

Vol. 33 Giuliana Garzone & Srikant Sarangi (eds)
Discourse, Ideology and Specialized Communication.
494 pages. 2007. ISBN 978-3-03910-888-6

Vol. 34 Giuliana Garzone & Cornelia Ilie (eds)
The Use of English in Institutional and Business Settings.
An Intercultural Perspective.
372 pages. 2007. ISBN 978-3-03910-889-3

Vol. 35 Vijay K. Bhatia & Maurizio Gotti (eds)
Explorations in Specialized Genres.
316 pages. 2006. ISBN 3-03910-995-2 · US-ISBN 0-8204-8372-9

Vol. 36 Heribert Picht (ed.)
Modern Approaches to Terminological Theories and Applications.
432 pages. 2006. ISBN 3-03911-156-6 · US-ISBN 0-8204-8380-X

Vol. 37 Anne Wagner & Sophie Cacciaguidi-Fahy (eds)
Legal Language and the Search for Clarity / Le langage juridique et la quête de clarté.
Practice and Tools / Pratiques et instruments.
487 pages. 2006. ISBN 3-03911-169-8 · US-ISBN 0-8204-8388-5

Vol. 38 Juan Carlos Palmer-Silveira, Miguel F. Ruiz-Garrido &
Inmaculada Fortanet-Gómez (eds)
Intercultural and International Business Communication:
Theory, Research and Teaching.
2nd edition. 343 pages. 2006, 2008. ISBN 978-3-03911-680-5

Vol. 39 Christiane Dalton-Puffer, Dieter Kastovsky, Nikolaus Ritt &
Herbert Schendl (eds)
Syntax, Style and Grammatical Norms: English from 1500–2000.
250 pages. 2006. ISBN 3-03911-181-7 · US-ISBN 0-8204-8394-X

Vol. 40 Marina Dossena & Irma Taavitsainen (eds)
Diachronic Perspectives on Domain-Specific English.
280 pages. 2006. ISBN 3-03910-176-0 · US-ISBN 0-8204-8391-5

Vol. 41 John Flowerdew & Maurizio Gotti (eds)
Studies in Specialized Discourse.
293 pages. 2006. ISBN 3-03911-178-7

Vol. 42 Ken Hyland & Marina Bondi (eds)
Academic Discourse Across Disciplines.
320 pages. 2006. ISBN 3-03911-183-3 · US-ISBN 0-8204-8396-6

Vol. 43 Paul Gillaerts & Philip Shaw (eds)
The Map and the Landscape: Norms and Practices in Genre.
256 pages. 2006. ISBN 3-03911-182-5 · US-ISBN 0-8204-8395-4

Vol. 44 Maurizio Gotti & Davide Giannoni (eds)
New Trends in Specialized Discourse Analysis.
301 pages. 2006. ISBN 3-03911-184-1 · US-ISBN 0-8204-8381-8

Vol. 45 Maurizio Gotti & Françoise Salager-Meyer (eds)
Advances in Medical Discourse Analysis: Oral and Written Contexts.
492 pages. 2006. ISBN 3-03911-185-X · US-ISBN 0-8204-8382-6

Vol. 46 Maurizio Gotti & Susan Šarcević (eds)
Insights into Specialized Translation.
396 pages. 2006. ISBN 3-03911-186-8 · US-ISBN 0-8204-8383-4

Vol. 47 Khurshid Ahmad & Margaret Rogers (eds)
Evidence-based LSP: Translation, Text and Terminology.
584 pages. 2007. ISBN 978-3-03911-187-9

Vol. 48 Hao Sun & Dániel Z. Kádár (eds)
It's the Dragon's Turn: Chinese Institutional Discourses.
262 pages. 2008. ISBN 978-3-03911-175-6

Vol. 49 Cristina Suárez-Gómez
Relativization in Early English (950-1250): the Position of Relative Clauses.
149 pages. 2006. ISBN 3-03911-203-1 · US-ISBN 0-8204-8904-2

Vol. 50 Maria Vittoria Calvi & Luisa Chierichetti (eds)
Nuevas tendencias en el discurso de especialidad.
319 pages. 2006. ISBN 978-3-03911-261-6

Vol. 51 Mari Carmen Campoy & María José Luzón (eds)
Spoken Corpora in Applied Linguistics.
274 pages. 2008. ISBN 978-3-03911-275-3

Vol. 52 Konrad Ehlich & Dorothee Heller (Hrsg.)
Die Wissenschaft und ihre Sprachen.
323 pages. 2006. ISBN 978-3-03911-272-2

Vol. 53 Jingyu Zhang
The Semantic Salience Hierarchy Model: The L2 Acquisition of Psych Predicates
273 pages. 2007. ISBN 978-3-03911-300-2

Vol. 54 Norman Fairclough, Giuseppina Cortese & Patrizia Ardizzone (eds)
Discourse and Contemporary Social Change.
555 pages. 2007. ISBN 978-3-03911-276-0

Vol. 55 Jan Engberg, Marianne Grove Ditlevsen, Peter Kastberg & Martin Stegu (eds)
New Directions in LSP Teaching.
331 pages. 2007. ISBN 978-3-03911-433-7

Vol. 56 Dorothee Heller & Konrad Ehlich (Hrsg.)
Studien zur Rechtskommunikation.
322 pages. 2007. ISBN 978-3-03911-436-8

Vol. 57 Forthcoming.

Vol. 58 Carmen Frehner
Email – SMS – MMS
294 pages. 2008. ISBN 978-3-03911-451-1

Vol. 59 Isabel Balteiro
The Directionality of Conversion in English: A Dia-Synchronic Study.
276 pages. 2007. ISBN 978-3-03911-241-8

Vol. 60 Maria Milagros Del Saz Rubio
English Discourse Markers of Reformulation.
237 pages. 2007. ISBN 978-3-03911-196-1

Vol. 61 Sally Burgess & Pedro Martín-Martín (eds)
English as an Additional Language in Research Publication and Communication.
259 pages. 2008. ISBN 978-3-03911-462-7

Vol. 62 Sandrine Onillon
Pratiques et représentations de l'écrit.
458 pages. 2008. ISBN 978-3-03911-464-1

Vol. 63 Hugo Bowles & Paul Seedhouse (eds)
Conversation Analysis and Language for Specific Purposes.
2nd edition. 337 pages. 2007, 2009. ISBN 978-3-0343-0045-2

Vol. 64 Vijay K. Bhatia, Christopher N. Candlin & Paola Evangelisti Allori (eds)
Language, Culture and the Law.
The Formulation of Legal Concepts across Systems and Cultures.
342 pages. 2008. ISBN 978-3-03911-470-2

Vol. 65 Jonathan Culpeper & Dániel Z. Kádár (eds)
Historical (Im)politeness.
300 pages. 2010. ISBN 978-3-03911-496-2

Vol. 66 Linda Lombardo (ed.)
Using Corpora to Learn about Language and Discourse.
237 pages. 2009. ISBN 978-3-03911-522-8

Vol. 67 Natsumi Wakamoto
Extroversion/Introversion in Foreign Language Learning.
Interactions with Learner Strategy Use.
159 pages. 2009. ISBN 978-3-03911-596-9

Vol. 68 Eva Alcón-Soler (ed.)
Learning How to Request in an Instructed Language Learning Context.
260 pages. 2008. ISBN 978-3-03911-601-0

Vol. 69 Domenico Pezzini
The Translation of Religious Texts in the Middle Ages.
428 pages. 2008. ISBN 978-3-03911-600-3

Vol. 70 Tomoko Tode
Effects of Frequency in Classroom Second Language Learning.
Quasi-experiment and stimulated-recall analysis.
195 pages. 2008. ISBN 978-3-03911-602-7

Vol. 71 Egor Tsedryk
Fusion symétrique et alternances ditransitives.
211 pages. 2009. ISBN 978-3-03911-609-6

Vol. 72 Cynthia J. Kellett Bidoli & Elana Ochse (eds)
English in International Deaf Communication.
444 pages. 2008. ISBN 978-3-03911-610-2

Vol. 73 Joan C. Beal, Carmela Nocera & Massimo Sturiale (eds)
Perspectives on Prescriptivism.
269 pages. 2008. ISBN 978-3-03911-632-4

Vol. 74 Carol Taylor Torsello, Katherine Ackerley & Erik Castello (eds)
Corpora for University Language Teachers.
308 pages. 2008. ISBN 978-3-03911-639-3

Vol. 75 María Luisa Pérez Cañado (ed.)
English Language Teaching in the European Credit Transfer System: Facing the Challenge.
251 pages. 2009. ISBN 978-3-03911-654-6

Vol. 76 Marina Dossena & Ingrid Tieken-Boon van Ostade (eds)
Studies in Late Modern English Correspondence. Methodology and Data.
291 pages. 2008. ISBN 978-3-03911-658-4

Vol. 77 Ingrid Tieken-Boon van Ostade & Wim van der Wurff (eds)
Current Issues in Late Modern English.
436 pages. 2009. ISBN 978-3-03911-660-7

Vol. 78 Marta Navarro Coy (ed.)
Practical Approaches to Foreign Language Teaching and Learning.
297 pages. 2009. ISBN 978-3-03911-661-4

Vol. 79 Qing Ma
Second Language Vocabulary Acquisition.
333 pages. 2009. ISBN 978-3-03911-666-9

Vol. 80 Martin Solly, Michelangelo Conoscenti & Sandra Campagna (eds)
Verbal/Visual Narrative Texts in Higher Education.
384 pages. 2008. ISBN 978-3-03911-672-0

Vol. 81 Meiko Matsumoto
From Simple Verbs to Periphrastic Expressions:
The Historical Development of Composite Predicates, Phrasal Verbs, and Related Constructions in English.
235 pages. 2008. ISBN 978-3-03911-675-1

Vol. 82 Melinda Dooly
Doing Diversity. Teachers' Construction of Their Classroom Reality.
180 pages. 2009. ISBN 978-3-03911-687-4

Vol. 83 Victoria Guillén-Nieto, Carmen Marimón-Llorca & Chelo Vargas-Sierra (eds)
Intercultural Business Communication and Simulation and Gaming Methodology.
392 pages. 2009. ISBN 978-3-03911-688-1

Vol. 84 Maria Grazia Guido
English as a Lingua Franca in Cross-cultural Immigration Domains.
285 pages. 2008. ISBN 978-3-03911-689-8

Vol. 85 Erik Castello
Text Complexity and Reading Comprehension Tests.
352 pages. 2008. ISBN 978-3-03911-717-8

Vol. 86 Maria-Lluisa Gea-Valor, Isabel García-Izquierdo & Maria-José Esteve (eds)
Linguistic and Translation Studies in Scientific Communication.
317 pages. 2010. ISBN 978-3-0343-0069-8

Vol. 87 Carmen Navarro, Rosa Mª Rodríguez Abella, Francesca Dalle Pezze & Renzo Miotti (eds)
La comunicación especializada.
355 pages. 2008. ISBN 978-3-03911-733-8

Vol. 88 Kiriko Sato
The Development from Case-Forms to Prepositional Constructions in Old English Prose.
231 pages. 2009. ISBN 978-3-03911-763-5

Vol. 89 Dorothee Heller (Hrsg.)
Formulierungsmuster in deutscher und italienischer Fachkommunikation.
Intra- und interlinguale Perspektiven.
315 pages. 2008. ISBN 978-3-03911-778-9

Vol. 90 Henning Bergenholtz, Sandro Nielsen & Sven Tarp (eds)
Lexicography at a Crossroads. Dictionaries and Encyclopedias Today,
Lexicographical Tools Tomorrow.
372 pages. 2009. ISBN 978-3-03911-799-4

Vol. 91 Manouchehr Moshtagh Khorasani
The Development of Controversies: From the Early Modern Period
to Online Discussion Forums.
317 pages. 2009. ISBN 978-3-3911-711-6

Vol. 92 María Luisa Carrió-Pastor (ed.)
Content and Language Integrated Learning: Cultural Diversity.
178 pages. 2009. ISBN 978-3-3911-818-2

Vol. 93 Roger Berry
Terminology in English Language Teaching: Nature and Use.
262 pages. 2010. ISBN 978-3-0343-0013-1

Vol. 94 Roberto Cagliero & Jennifer Jenkins (eds)
Discourses, Communities, and Global Englishes
240 pages. 2010. ISBN 978-3-0343-0012-4

Vol. 95 Facchinetti Roberta, Crystal David, Seidlhofer Barbara (eds)
From International to Local English – And Back Again.
268 pages. 2010. ISBN 978-3-0343-0011-7

Vol. 96 Cesare Gagliardi & Alan Maley (eds)
EIL, ELF, Global English: Teaching and Learning Issues
376 pages. 2010. ISBN 978-3-0343-0010-0

Vol. 97 Sylvie Hancil (ed.)
The Role of Prosody in Affective Speech.
403 pages. 2009. ISBN 978-3-03911-696-6

Vol. 98 Marina Dossena & Roger Lass (eds)
Studies in English and European Historical Dialectology.
257 pages. 2009. ISBN 978-3-0343-0024-7

Vol. 99 Christine Béal
Les interactions quotidiennes en français et en anglais.
De l'approche comparative à l'analyse des situations interculturelles.
424 pages. 2010. ISBN 978-3-0343-0027-8

Vol. 100 Maurizio Gotti (ed.)
Commonality and Individuality in Academic Discourse.
398 pages. 2009. ISBN 978-3-0343-0023-0

Vol. 101 Javier E. Díaz Vera & Rosario Caballero (eds)
Textual Healing: Studies in Medieval English Medical,
Scientific and Technical Texts.
213 pages. 2009. ISBN 978-3-03911-822-9

Vol. 102 Nuria Edo Marzá
The Specialised Lexicographical Approach: A Step further in Dictionary-making.
316 pages. 2009. ISBN 978-3-0343-0043-8

Vol. 103 Carlos Prado-Alonso, Lidia Gómez-García, Iria Pastor-Gómez &
David Tizón-Couto (eds)
New Trends and Methodologies in Applied English Language Research.
Diachronic, Diatopic and Contrastive Studies.
348 pages. 2009. ISBN 978-3-0343-0046-9

Vol. 104 Françoise Salager-Meyer & Beverly A. Lewin
Crossed Words: Criticism in Scholarly Writing?
371 pages. 2011. ISBN 978-3-0343-0049-0.

Vol. 105 Javier Ruano-García
Early Modern Northern English Lexis: A Literary Corpus-Based Study.
611 pages. 2010. ISBN 978-3-0343-0058-2

Vol. 106 Rafael Monroy-Casas
Systems for the Phonetic Transcription of English: Theory and Texts.
280 pages. 2011. ISBN 978-3-0343-0059-9

Vol. 107 Nicola T. Owtram
The Pragmatics of Academic Writing.
A Relevance Approach to the Analysis of Research Article Introductions.
311 pages. 2009. ISBN 978-3-0343-0060-5

Vol. 108 Yolanda Ruiz de Zarobe, Juan Manuel Sierra &
Francisco Gallardo del Puerto (eds)
Content and Foreign Language Integrated Learning.
Contributions to Multilingualism in European Contexts
343 pages. 2011. ISBN 978-3-0343-0074-2

Vol. 109 Ángeles Linde López & Rosalía Crespo Jiménez (eds)
Professional English in the European context: The EHEA challenge.
374 pages. 2010. ISBN 978-3-0343-0088-9

Vol. 110 Rosalía Rodríguez-Vázquez
The Rhythm of Speech, Verse and Vocal Music: A New Theory.
394 pages. 2010. ISBN 978-3-0343-0309-5

Vol. 111 Anastasios Tsangalidis & Roberta Facchinetti (eds)
Studies on English Modality. In Honour of Frank Palmer.
392 pages. 2009. ISBN 978-3-0343-0310-1

Vol. 112 Forthcoming.

Vol. 113 Mihhail Lotman & Maria-Kristiina Lotman (eds)
Frontiers in Comparative Prosody. In memoriam: Mikhail Gasparov.
426 pages. 2011. ISBN 978-3-0343-0373-6

Vol. 114 Merja Kytö, John Scahill & Harumi Tanabe (eds)
Language Change and Variation from Old English to Late Modern English.
A Festschrift for Minoji Akimoto
422 pages. 2010. ISBN 978-3-0343-0372-9

Vol. 115 Giuliana Garzone & Paola Catenaccio (eds)
Identities across Media and Modes: Discursive Perspectives.
379 pages. 2009. ISBN 978-3-0343-0386-6

Vol. 116 Elena Landone
Los marcadores del discurso y cortesía verbal en español.
390 pages. 2010. ISBN 978-3-0343-0413-9

Vol. 117 Maurizio Gotti & Christopher Williams (eds)
Legal Discourse across Languages and Cultures.
339 pages. 2010. ISBN 978-3-0343-0425-2

Vol. 118 David Hirsh
Academic Vocabulary in Context.
217 pages. 2010. ISBN 978-3-0343-0426-9

Vol. 119 Yvonne Dröschel
Lingua Franca English. The Role of Simplification and Transfer.
358 pages. 2011. ISBN 978-3-0343-0432-0

Vol. 120 Tengku Sepora Tengku Mahadi, Helia Vaezian & Mahmoud Akbari
Corpora in Translation. A Practical Guide.
135 pages. 2010. ISBN 978-3-0343-0434-4

Vol. 121 Davide Simone Giannoni & Celina Frade (eds)
Researching Language and the Law. Textual Features and Translation Issues.
278 pages. 2010. ISBN 978-3-0343-0443-6

Vol. 122 Daniel Madrid & Stephen Hughes (eds)
Studies in Bilingual Education.
472 pages. 2011. ISBN 978-3-0343-0474-0

Vol. 123 Vijay K. Bhatia, Christopher N. Candlin & Maurizio Gotti (eds)
The Discourses of Dispute Resolution.
290 pages. 2010. ISBN 978-3-0343-0476-4

Vol. 124 Davide Simone Giannoni
Mapping Academic Values in the Disciplines. A Corpus-Based Approach.
288 pages. 2010. ISBN 978-3-0343-0488-7

Vol. 125 Giuliana Garzone & James Archibald (eds)
Discourse, Identities and Roles in Specialized Communication.
419 pages. 2010. ISBN 978-3-0343-0494-8

Vol. 126 Iria Pastor-Gómez
The Status and Development of N+N Sequences in
Contemporary English Noun Phrases.
216 pages. 2011. ISBN 978-3-0343-0534-1

Vol. 127 Carlos Prado-Alonso
Full-verb Inversion in Written and Spoken English.
261 pages. 2011. ISBN 978-3-0343-0535-8

Vol. 128 Tony Harris & María Moreno Jaén (eds)
Corpus Linguistics in Language Teaching.
214 pages. 2010. ISBN 978-3-0343-0524-2

Vol. 129 Tetsuji Oda & Hiroyuki Eto (eds)
Multiple Perspectives on English Philology and History of Linguistics.
A Festschrift for Shoichi Watanabe on his 80th Birthday.
378 pages. 2010. ISBN 978-3-0343-0480-1

Vol. 130 Luisa Chierichetti & Giovanni Garofalo (eds)
Lengua y Derecho: líneas de investigación interdisciplinaria.
283 pages. 2010. 978-3-0343-0463-4

Vol. 131 Paola Evangelisti Allori & Giuliana Garzone (eds)
Discourse, Identities and Genres in Corporate Communication.
Sponsorship, Advertising and Organizational Communication.
324 pages. 2011. 978-3-0343-0591-4

Vol. 132 Leyre Ruiz de Zarobe & Yolanda Ruiz de Zarobe (eds)
Speech Acts and Politeness across Languages and Cultures.
402 pages. 2012. 978-3-0343-0611-9

Vol. 133 Thomas Christiansen
Cohesion: A Discourse Perspective.
387 pages. 2011. 978-3-0343-0619-5

Vol. 134 Giuliana Garzone & Maurizio Gotti
Discourse, Communication and the Enterprise. Genres and Trends.
451 pages. 2011. ISBN 978-3-0343-0620-1

Vol. 135 Zsuzsa Hoffmann
Ways of the World's Words.
Language Contact in the Age of Globalization.
334 pages 2011. ISBN 978-3-0343-0673-7

Vol. 136 Cecilia Varcasia (ed.)
Becoming Multilingual.
Language Learning and Language Policy between Attitudes and Identities.
213 pages. 2011. ISBN 978-3-0343-0687-5

Vol. 137 Susy Macqueen
The Emergence of Patterns in Second Language Writing.
A Sociocognitive Exploration of Lexical Trails.
325 pages. 2012. ISBN 978-3-0343-1010-9

Vol. 138 Maria Vittoria Calvi & Giovanna Mapelli (eds)
La lengua del turismo. Géneros discursivos y terminología.
365 pages. 2011. ISBN 978-3-0343-1011-6

Vol. 139 Ken Lau
Learning to Become a Professional in a Textually-Mediated World.
A Text-Oriented Study of Placement Practices.
261 pages. 2012. ISBN 978-3-0343-1016-1

Vol. 140 Forthcoming.

Vol. 141 Edith Esch & Martin Solly (eds)
The Sociolinguistics of Language Education in International Contexts.
263 pages. 2012. ISBN 978-3-0343-1009-3

Vol. 142–143 Forthcoming.

Vol. 144 Margrethe Petersen & Jan Engberg (eds)
Current Trends in LSP Research. Aims and Methods.
323 pages. 2011. ISBN 978-3-0343-1054-3

Vol. 145 David Tizón-Couto, Beatriz Tizón-Couto, Iria Pastor-Gómez & Paula Rodríguez-Puente (eds)
New Trends and Methodologies in Applied English Language Research II.
Studies in Language Variation, Meaning and Learning.
283 pages. 2012. ISBN 978-3-0343-1061-1

Vol. 146 Rita Salvi & Hiromasa Tanaka (eds)
Intercultural Interactions in Business and Management.
306 pages. 2011. ISBN 978-3-0343-1039-0

Vol. 147 Francesco Straniero Sergio & Caterina Falbo (eds)
Breaking Ground in Corpus-based Interpreting Studies.
254 pages. 2012. ISBN 978-3-0343-1071-0

Vol. 148 Forthcoming.

Vol. 149 Vijay K. Bhatia & Paola Evangelisti Allori (eds)
Discourse and Identity in the Professions. Legal, Corporate and Institutional Citizenship.
352 pages. 2011. ISBN 978-3-0343-1079-6

Vol. 150 Maurizio Gotti (ed.)
Academic Identity Traits. A Corpus-Based Investigation.
363 pages. 2012. ISBN 978-3-0343-1141-0

Vol. 151 Priscilla Heynderickx, Sylvain Dieltjens, Geert Jacobs, Paul Gillaerts & Elizabeth de Groot (eds)
The Language Factor in International Business.
New Perspectives on Research, Teaching and Practice.
320 pages. 2012. ISBN 978-3-0343-1090-1

Vol. 152 Paul Gillaerts, Elizabeth de Groot, Sylvain Dieltjens, Priscilla Heynderickx & Geert Jacobs (eds)
Researching Discourse in Business Genres. Cases and Corpora.
215 pages. 2012. ISBN 978-3-0343-1092-5

Vol. 153 Yongyan Zheng
Dynamic Vocabulary Development in a Foreign Language.
262 pages. 2012. ISBN 978-3-0343-1106-9

Vol. 154 Forthcoming.

Vol. 155 David Hirsh (ed.)
Current Perspectives in Second Language Vocabulary Research.
180 pages. 2012. ISBN 978-3-0343-1108-3

Vol. 156–157 Forthcoming.

Vol. 158 Bárbara Eizaga Rebollar (ed.)
Studies in Linguistics and Cognition.
301 pages. 2012. ISBN 978-3-0343-1138-0

Vol. 159 Giuliana Garzone, Paola Catenaccio, Chiara Degano (eds)
Genre Change in the Contemporary World. Short-term Diachronic Perspectives.
329 pages. 2012. ISBN 978-3-0343-1214-1

Vol. 160 Carol Berkenkotter, Vijay K. Bhatia & Maurizio Gotti (eds)
Insights into Academic Genres.
468 pages. 2012. ISBN 978-3-0343-1211-0

Vol. 161 Forthcoming.

Vol. 162 Patrizia Anesa
Jury Trials and the Popularization of Legal Language. A Discourse Analytical Approach.
247 pages. 2012. ISBN 978-3-0343-1231-8